STUDIES

IN THE

SOCIAL ASPECTS

OF THE

DEPRESSION

Studies in the Social Aspects of the Depression

Advisory Editor: *ALEX BASKIN*

State University of New York at Stony Brook

RESEARCH MEMORANDUM ON SOCIAL ASPECTS OF HEALTH IN THE DEPRESSION

By SELWYN D. COLLINS

and CLARK TIBBITTS

ARNO PRESS

A NEW YORK TIMES COMPANY

Reprint Edition 1972 by Arno Press Inc.

Reprinted from a copy in The Newark Public Library

LC# 72-162846
ISBN 0-405-00849-X

Studies in the Social Aspects of the Depression
ISBN for complete set: 0-405-00840-6
See last pages of this volume for titles.

Manufactured in the United States of America

Preface to the New Edition

THOUGH FORTUNE MAGAZINE HAS CLAIMED that no one starved during the Depression—a point which has been challenged in some quarters—there can be little doubt that there was a close relationship between economic status and health care. With the loss of jobs and the decline in family income one could expect that there would be a decline in the medical care available for those without funds. Studies described by Collins and Tibbitts in this work substantiate statistically what was suspected intellectually. Physicians in California acknowledged that their volume of paid practice decreased by 9 percent between 1929 and 1934. In fact, a study conducted by the Federal Emergency Relief Administration revealed that among those seeking public assistance there was a small contingent of doctors.

Living as we do in an era of Medicare and Medicaid, social security and workman's compensation, Blue Cross and Blue Shield and a host of other health insurance plans, it is difficult to recall how few protections there once were to help a family in troubled times. While conditions are improved in our decade, it would be misleading to suggest that full and equitable medical care is available to all. The facts indicate the contrary; there is a substantial segment of the population who still stand alone, unshielded, unprotected and vulnerable to the ravages of illness and accident.

The authors have examined the numerous health related issues of the 1930's and have identified those themes and topics which require further research and investigation. There is still much in this work to suggest initiatives and directions that might help enrich the quality of life in the 1970's and beyond.

Alex Baskin
Stony Brook, New York, 1971

BULLETIN 36

1937

RESEARCH MEMORANDUM ON SOCIAL ASPECTS OF HEALTH IN THE DEPRESSION

By SELWYN D. COLLINS
Principal Statistician in charge of Statistical Investigations
United States Public Health Service

and CLARK TIBBITTS
Chairman Health Inventory Operating Council
United States Public Health Service

with the Assistance of
ARCH B. CLARK AND ELEANOR L. RICHIE

PREPARED UNDER THE DIRECTION OF THE
COMMITTEE ON STUDIES IN SOCIAL
ASPECTS OF THE DEPRESSION

SOCIAL SCIENCE RESEARCH COUNCIL
230 PARK AVENUE NEW YORK NY

The Social Science Research Council was organized in 1923 and formally incorporated in 1924, composed of representatives chosen from the seven constituent societies and from time to time from related disciplines such as law, geography, psychiatry, medicine, and others. It is the purpose of the Council to plan, foster, promote, and develop research in the social field.

CONSTITUENT ORGANIZATIONS

American Anthropological Association

American Economic Association

American Historical Association

American Political Science Association

American Psychological Association

American Sociological Society

American Statistical Association

FOREWORD

By the Committee on Studies in Social Aspects of the Depression

THIS monograph on research pertaining to the social aspects of health in the depression is one of a series of thirteen sponsored by the Social Science Research Council to stimulate the study of depression effects on various social institutions. The full list of titles is on page ii.

The depression of the early 1930's was like the explosion of a bomb dropped in the midst of society. All the major social institutions, such as the government, family, church, and school, obviously were profoundly affected and the repercussions were so far reaching that scarcely any type of human activity was untouched.

It would be valuable to have assembled the vast record of influence of this economic depression on society. Such a record would constitute an especially important preparation for meeting the shock of the next depression, if and when it comes. The facts about the impact of the depression on social life have been only partially recorded. Theories must be discussed and explored now, if much of the information to test them is not to be lost amid ephemeral sources.

The field is so broad that selection has been necessary. In keeping with its mandate from the Social Science Research Council, the Committee sponsored no studies of an exclusively economic or political nature. The subjects chosen for inclusion in the series were limited in number by available resources. The final selection was made by the Committee from a much larger number of proposed subjects, on the basis of social importance and available personnel.

Although the monographs clearly reveal a uniformity of goal, they differ in the manner in which the various authors sought to attain that goal. This is a consequence of the Committee's belief that the promotion of research could best be served by not imposing rigid restrictions on the organization of materials by the contributors. It is felt that the encouraged freedom in approach and organization has resulted in the enrichment of the individual reports and of the series as a whole.

A common goal without rigidity in procedure was secured by requesting each author to examine critically the literature on the depression for the purpose of locating existing data and interpretations already reasonably well established, of discovering the more serious inadequacies in information, and of formulating research problems feasible for study. He was not expected to do this research himself. Nor was he expected to compile a full and systematically treated record of the depression as experienced in his field. Nevertheless, in indicating the new research which is needed, the writers found it necessary to report to some extent on what is known. These volumes actually contain much information on the social influences of the depression, in addition to their analyses of pressing research questions.

The undertaking was under the staff direction of Dr. Samuel A. Stouffer, who worked under the restrictions of a short time limit in order that prompt publication might be assured. He was assisted by Mr. Philip M. Hauser and Mr. A. J. Jaffe. The Committee wishes to express appreciation to the authors, who contributed their time and effort without remuneration, and to the many other individuals who generously lent aid and materials.

William F. Ogburn Chairman
Shelby M. Harrison
Malcolm M. Willey

PREFACE

IN CONCLUDING his monograph on *Health and Environment,* Sydenstricker stated ". . . there is a large body of evidence to show that many of the variations among individuals with respect to mortality, morbidity, and physical impairment are due chiefly to environmental factors. The occurrence of certain diseases, such as simple goiter, is almost entirely determined by geographic location; urban life, unless counteracted by specific preventive efforts, is less favorable for survival and healthful living than rural life; the prevalence of infectious diseases is determined largely by such conditions as water supplies, facilities for sanitation, opportunity for contact and for immunization—whether 'natural' or artificial; occupation and occupational environment appear to be important factors in determining the mortality from tuberculosis in adult ages; diet, as determined by food supplies, dietary habits, and economic status, is an important environmental factor in health; the ability to purchase favorable dwelling facilities, medical care, proper food, sufficient rest and leisure, is a factor arising out of a complex set of environmental conditions; social customs, traditions, attitudes—all of these have some effect upon health."[1] If one accepts the hypothesis that environment as well as heredity plays a large rôle in the health of a population, then the present monograph definitely belongs in a series of memoranda having to do with social aspects of the depression.

One might raise the question whether the state of knowledge and the availability of data warrant so prominent a considera-

[1] Sydenstricker, Edgar. *Health and Environment.* A Committee on Recent Social Trends Monograph. New York: McGraw-Hill Book Co. 1933. Pp. 207-208

tion of the social aspects of health in relation to the depression. The health of an individual or of a group is in itself difficult to measure. When one undertakes to determine the relationship between environment and health, he finds himself facing an even more complex problem, for the relationship between health and one environmental factor must be studied while the influence of heredity and of numerous other phases of the environment are held constant. Because of these complexities data are difficult to find; particularly data concerning the status of health over a period of time. It has been necessary to suggest studies of health and of various phases of the environment independently of one another. Such studies would be made often on the basis of hypotheses concerning the relationship between environment and health.

In the same manner it has been difficult to provide a setting for new research through a review of what is already known about the effect of the depression on social aspects of health. Many presumable consequences of the depression are, therefore, cited as possibilities rather than facts.

Nevertheless, the need for continued research in the field is very great, for good health is an important possession in times of prosperity as well as depression. When depression strikes it may tend to select the unhealthy as candidates for public assistance. It may contribute to the immediate ill health of others, thus delaying their chances of re-employment. Finally it may impair permanently the health of still others through undernourishment, postponement of needed medical attention, and restriction of the activities of health agencies and professional persons. Much research is needed to point the way toward a greater conservation of health and a more efficient use of health services during future depressions and in times of prosperity.

The references cited in the following pages are not limited to those having a bearing on the depression or even to those relating to economic status and health. It was not possible in this monograph to summarize in any great detail the research that has been mentioned, in large part, as a general background for studies of a more economic nature. In lieu of such a summary, frequent and often general references have been inserted so that the student who is interested in any specific suggested study may, without too much research, review the background and arrive at his own conclusion as to whether the suggested study is worthy of further effort.

No attempt has been made to assemble a complete bibliography in any of the fields covered. The references cited are illustrative of the kinds of data available but in many instances contain further references which will lead to a more complete survey of the field. In order to conserve space the references have not been collected in a formal bibliography, but are cited in full detail in the footnotes.

In the preparation of the monograph the authors have profited from the assistance and advice of several members of the staff of the United States Public Health Service and of the National Health Inventory as well as of other individuals. Miss Richie and Mr. Clark abstracted a great deal of material and prepared early drafts of several sections of the report. Mr. Clark contributed the sections on occupational environment and changes in social environment. Kenneth H. McGill contributed the sections on housing and mobility. David Clingersmith put the footnotes in their final form, read the proof, and assumed most of the responsibility for preparing the index. Many valuable criticisms of the manuscript were given by Joseph W. Mountin, G. St. J. Perrott, A. J. Borowski, Margaret L. Plumley, Nathan Sinai, Rollo H. Britten, Michael Davis, Dorothy Wiehl, and the Director and the Committee at whose request the work was undertaken.

Washington D.C. S D C *and* C T
July 15 1937

CONTENTS

FOREWORD v

PREFACE vii

I THE PROBLEM 1

Economic Status and Health 1
Sickness, the Health Environment, and the Receipt of Care 3
Problems of Method 5

II MEASURES OF HEALTH 9

Infant, Neo-Natal, and Maternal Mortality and Stillbirths 12
Neo-Natal and Other Infant Mortality 13
Maternal Mortality 16
Stillbirths, Miscarriages, and Abortions 18
Studies Suggested 19

General Mortality, Particularly of Adult Males of the Working Ages 25
Mortality from Various Diseases 26
Mortality from Accidents 28
Mortality from Suicide 28
Studies Suggested 29

Morbidity or Sickness Rates 32
General Population 32
Members of Sick Benefit Associations and Other Insured Persons 40
School Children 42
Hospitalized Illness 46
Reportable Diseases 47
Venereal Diseases 48
Studies Suggested 49

Accidents 57

Nutrition, Height, and Weight 59
Underweight 61
Nutrition as Judged by Clinical Manifestations 64
Derived Effects of the Depression 66
Studies Suggested 66

III ENVIRONMENT AND HEALTH 73

Minimum Income for Subsistence 74
Determination of Minimum Subsistence Standards 75
Incomes Below Subsistence Standard 77
Adequacy of Incomes of Relief Families 81
Summary 85

Consumption 86
Housing 86
Clothing and Household Necessities 92
Food 94
Summary 96

Occupational Environment 96
The Changing Environment 97
Working Conditions and the Depression 100
Studies Suggested 102

Changes in Social Environment 105
The Family 106
Education and Communication 107
Social Work Other Than Relief 108
Mobility 109

IV THE PREVENTION AND TREATMENT OF ILLNESS 113

The Volume of Preventive Services 114
Environmental Control 114
Health Education 125
Direct Services 127
Summary 133
Studies Suggested 134

The Receipt of Medical Care 140
Provision of Care 142
Medical Care and Economic Status 150
Studies Suggested 157

Organization for Medical Care 161
Medical Practice and the Depression 161
The Community Takes a Hand 166
Summary 170
Studies Suggested 170

V SUMMARY 175

INDEX 183

Chapter I

The Problem

THE relation of economic status to health has been a favorite topic for research in the field of public health. A most impressive evidence of interest in this problem is a study made in 1918[1] by Dr. Charles V. Chapin, then dean of American health officers. Dr. Chapin analyzed mortality among taxpayers and non-taxpayers in Providence (R.I.) for the year 1865 when an income tax law was in force in that city. He had hoped to make such a study with data derived from returns under the Federal Income Tax law of 1913; but, when he found that these data were not available to him in the necessary detail (individual's name), his great interest in the subject took him back to the records of 1865 to make the study.

ECONOMIC STATUS AND HEALTH

Although the "effect of economic status upon health" is the phase of the "relation between economic status and health" which is most frequently discussed and which is our present concern, the relationship is not one of simple cause and effect. One need consider only his personal knowledge of self-supporting families who have been precipitated into the impoverished classes by some catastrophic chronic illness of the chief breadwinner to be convinced that sickness may be a cause as well as a result of poverty. It appears to be the consensus of scientific

[1] Chapin, Charles V. "Deaths among Taxpayers and Non-Taxpayers, Income Tax, Providence (R.I.), 1865." *Papers of Charles V. Chapin*. New York: The Commonwealth Fund. 1934. Pp. 217-228

opinion, however, that poverty gives rise to sickness more frequently than sickness gives rise to poverty.[2]

It is elemental, of course, that income as such has no effect upon illness save as it determines or influences the mode of living of the recipient and his family, including the adequacy of medical care which is received. Standards of living and health are also functions of personal characteristics such as intelligence, education, resourcefulness, and initiative. This fact undoubtedly contributes materially to the normally high correlation between income and health, since the competitive process tends, at least, to place the individuals who demonstrate a high degree of development of these characteristics in the upper income groups. That is, under ordinary economic conditions, the poor may average lower in education, intelligence, and other characteristics than the higher income groups, and hence may be less able than are members of the comfortable classes to use to advantage what income they have.

For a few diseases, notably pellagra, studies have revealed the exact method by which the mode of living of the impoverished individual leads to the disease,[3] but in the great majority of instances the deficiencies in the standard of living that favor the development of the disease have not been determined. The absence of this information may indicate that concentrated research has not been done, or that the relationship is so complex that it cannot be separated into its components. For example, intelligence, education, resourcefulness, and initiative vary among individuals at any income level, as well as among economic classes, and this variation makes the nature of the re-

---

[2] This has been discussed by Perrott, G. St. J. and Sydenstricker, E. "Causal and Selective Factors in Sickness." *American Journal of Sociology*. Vol. 40. No. 6. May 1935

[3] See Goldberger, Joseph, Wheeler, G. A., Sydenstricker, Edgar, King, Wilford I. *et al.* "A Study of Endemic Pellagra in Some Cotton-Mill Villages of South Carolina." *Hygienic Laboratory Bulletin*. No. 153. Washington: Government Printing Office. January 1929

lationship between illness and economic status more difficult to determine.

It is pertinent, nevertheless, to inquire about changes in health and in the receipt of preventive services and treatment that may have resulted from depression declines in income and the consequent lowering of living standards of large sections of the population.[4]

SICKNESS, THE HEALTH ENVIRONMENT, AND THE RECEIPT OF CARE

The purpose of this memorandum is to discuss probable changes and to suggest social research in the fields of: (1) health; (2) certain aspects of the health environment; and (3) the practices and institutions for the prevention and treatment of illness. The three main chapters of the monograph represent these three topics.

Health.— The first topic has to do with measures of health and changes in health during the depression as compared with preceding years in such terms as the frequency of sickness and death. This is, of course, the chief interest in any consideration of the effects of the depression on health.

The Health Environment.— The second topic has to do with depression changes in environmental conditions that may reasonably be assumed to affect health, including incomes as a measure of the material standard of living, occupational and home surroundings, and certain social factors. While to include these subjects may not seem altogether warranted in view of the lack of knowledge concerning the nature of their influence on well-being, the decision to do so is entirely in accord with current statements of the scope of public health.

The history of public health has been one of continuously

[4] A discussion of the consequences of the depression on various social factors is found in Woytinsky, Wladimir. *The Social Consequences of the Economic Depression*. Studies and Reports. Series C. Employment and Unemployment. No. 21. Geneva: International Labour Office (League of Nations). 1936

widening objectives. Its early efforts were directed to furthering the sanitation of the physical environment through such activities as inspection of plumbing, garbage and refuse collection, sewage disposal, and provision of potable water supplies. Having established these functions, health officers struck out into the new field of communicable disease control. While this problem is by no means solved, there has been marked success with respect to diseases like smallpox, diphtheria, tuberculosis, infantile diarrhea, and typhoid fever. Syphilis is not under control but the lines of action leading to that end are fairly well charted, and public health is turning its attention to diseases like cancer, diabetes, cardiac conditions, and to maternal mortality. An important item in the control of these problems turns upon the question of making medical services available to those who are not now able to obtain them. From raising the general level of the receipt of medical services, it is only a short step in the public health program to recognition of the significance of nutrition, housing, working conditions, and recreation. While the relation between several of these items and health is still obscure, there is little doubt but that the physical and mental well-being of the population is largely dependent upon them. Hence we find control of these factors discussed in nearly identical statements by such leaders in public health as the late Edgar Sydenstricker,[5] Dr. Thomas Parran, the Surgeon General of the United States Public Health Service,[6] and Professor C.-E. A. Winslow of the Yale University School of Medicine.[7] Depression changes among these items are therefore important in themselves even though they do not measure actual changes in health.

[5] Sydenstricker, Edgar. "Next Steps in Public Health." *Next Steps in Public Health.* New York: Milbank Memorial Fund. 1936. Pp. 13-34

[6] Parran, Thomas. "Reporting Progress." Presidential address. *American Journal of Public Health.* 26:1071-1076. No. 11. November 1936

[7] Winslow, C.-E. A. "Next Steps: A Review of the Conference." *Next Steps in Public Health.* New York: Milbank Memorial Fund. 1936. Pp. 71-77

The Prevention and Treatment of Illness.—The third topic has to do with changes in the distribution of medical treatment expressed in terms of the amount of care received and in terms of changes in the purposes, programs, volume of service rendered, expenditures, and number of persons employed in private medical practice and among health departments, hospitals, clinics, and other agencies which are part of the organization for the prevention and care of sickness. While these activities are closely related to the volume of illness, changes in them cannot be interpreted as measures of health during the depression. They are included because the depression may have initiated or hastened changes that are important for the future health of the population.

PROBLEMS OF METHOD

The first topic, on measures of health per se, lends itself to treatment in a direct statistical manner. The second and third topics are less susceptible of quantitative analysis because of lack of data and because the relationships between health and the various items considered are not always clear cut. The nature of measures of health, the limitations of available methods of collecting data, and the shortcomings of existing records offer several methodological problems which may be raised appropriately at this point. First, studies of the health of low income classes during periods of free business activity may not be appropriate for estimating the effect of the depression upon the health of families whose incomes were curtailed. It is reasonable to suppose that those who were thrown from the higher to the lower income groups had more initiative, resourcefulness, education, and intelligence than the habitually poor groups even though the new poor came largely from the margin of the comfortable classes. Consequently, one would hardly expect to find the same difference in the health of the new poor prior to and during the depression as is commonly found in the health of the comfortable and the poor in a normal business period. On the other hand, the necessity of making quick adjustment to the poverty which comes with depression may overtax the superior abilities and resourcefulness of the new poor and thus increase their illness and death rates even above those of the chronic poor.

Records of Sickness and Care.—Records of health agencies, hospitals, clinics, and physicians constitute a potential source of chronological data showing changes in the amount of sickness and in the volume of medical care, but reporting in this field is relatively new and the data frequently lack completeness and definitiveness.

The incidence of births and deaths is quite fully registered but not all of the facts pertaining to them are well reported. Most communities require reporting of communicable diseases but the list of diseases for which reports are required and the completeness of reporting vary strikingly from one place to another. Reports of the venereal diseases, tuberculosis, and pneumonia are particularly incomplete, and most diseases fall entirely outside the reportable category, so that official records offer little in the way of adequate data for the student of trends. Hence in order to obtain anything like an adequate statement of the amount of sickness experienced it would be necessary to go to the records of the physicians and institutions providing treatment.

This method of determining the amount of sickness is seriously limited by its failure to include records of illnesses that do not receive treatment, and it may well be imagined that the proportion of all sickness that is treated varies between prosperous and depression periods. It suffers also from the limitation that, with some exceptions, it provides sickness data for only the total population and not for the different income or racial groups, or other classes that were probably not all affected in the same way by the depression.

Records of health agencies and practitioners, if they were complete, might yield fairly good estimates of the total volume of services rendered, but they, too, would fail to establish the important differentials among various segments of the population. These data would also be deficient in that they could not be related to the total volume of sickness experienced.

House-to-House Canvasses.—House-to-house canvasses have been extremely useful in obtaining a picture of the incidence and cause of illness for persons of different age, sex, and income groups, together with salient facts about each illness, such as its duration, in terms of days lost from work or days in bed, the amounts and types of medical services received, in terms of the kind of attendant, the number of calls by the doctor, the days and visits of nursing care, and the number of days in the hospital. This method obviously overcomes the difficulty of determining rates of sickness and the receipt of care among income or other groups and of relating the volume of treatment to the amount of sickness. It presents a major problem, however, in that the reporting is dependent upon the memory and knowledge of the families canvassed. So far as memory is concerned one should be able to obtain through careful family interviews a complete record for many years back of illness of such serious types as typhoid fever, appendicitis, pneumonia, tuberculosis, cancer, apoplexy and insanity. Also, a mother is usually able to give a fairly complete history of the communicable diseases of her children of the school and preschool ages. Minor illnesses, however, such as colds, influenza, laryngitis, tonsillitis and digestive upsets, are easily forgotten and many times are not diagnosed, yet they are just as important as the more serious conditions for the purpose of studying the effects of depressions. Furthermore, deaths and commitments to institutions tend to break up families, particularly when the patient is one of the family heads. As a result, the existing families that are surveyed do not report these illnesses because they did not occur within the present family. These same factors are equally operative in connection with efforts to measure the receipt of medical care. Their importance is so great that it is not deemed feasible to attempt to collect records save for serious conditions or for major items of treatment for a period of more than one year prior to the date of the survey.

More specific reference to these problems is made in the discussion and in connection with the suggestions for research for each of the three topics in the following chapters.

Chapter II

Measures of Health

MEASURES of health may be considered in relation to the individual and in relation to a given population group. One might think of these as two phases of the same process, that is, that one would (1) determine physical fitness of individuals and classify them according to some standard and (2) relate these classes to the total population under consideration. If such a plan were feasible and practicable, it would no doubt be the most desirable method, for it could take into account in the measure of physical fitness or unfitness of the group any degree of fitness or unfitness of the individual that seemed desirable.

Actually, the assessment of the physical fitness of a given individual is a thing that has never been developed to a rule-of-thumb process. If one postulates the use and assistance of every possible technique including physical examinations, laboratory tests and sickness records, one still has only an imperfect measure of the health and viability of an individual.

The problem of assessing the physical fitness of an individual has been attacked by many people; among the physical standards set up are height-weight-age tables;[1] various other measures to

---

[1] See Baldwin, B. T. and Wood, T. D. *Tables for Boys and Girls of School Age.* Supplement to July 1923 issue of *Mother and Child.* Washington: American Child Health Association; and "Height-Weight-Age Tables for Adults" prepared by life insurance companies and reproduced by Britten, R. H. and Thompson, L. R. *A Health Study of Ten Thousand Male Industrial Workers.* Public Health Bulletin No. 162. June 1926. Washington: Government Printing Office. 1926. P. 160

supplement height and weight, such as sitting height, chest measurements, vital capacity;[2] and measurements of subcutaneous tissue and fat to supplement chest, hip and arm measurements.[3] In the feeding of world war refugees one of these methods of assessing physical fitness was used extensively to pick out undernourished children.[4] One could add many references to these attempts to establish a rule-of-thumb system to replace the more laborious and less standard procedure of physical examination and assessment of the health or "nutrition" of the individual on the basis of clinical evidence. Some of these physical standards have been studied and compared with physical examination findings to determine their ability to select the undernourished and the unfit.[5] The general conclusion of all who have studied them is that all methods that have thus far been suggested for assessing the physical fitness of individuals are far from perfect.

Even a satisfactory standard of physical fitness would be almost impossible to use in the study of health and economic depression because it would be necessary to examine and measure in a uniform way many individuals in the course of the depres-

---

[2] See Dreyer, G. *The Assessment of Physical Fitness.* New York and London: Cassell and Co., Ltd. 1920; Pirquet, C. *An Outline of the Pirquet System of Nutrition.* Philadelphia and London: W. B. Saunders Co. 1922

[3] See Franzen, R. and Palmer, G. T. *The ACH Index of Nutritional Status.* New York: American Child Health Association. 1934; *Physical Measures of Growth and Nutrition.* New York: American Child Health Association. 1929; and Taylor, C. K. *Physical Standards for Boys and Girls.* Orange, N.J.: The Academy Press. 1922

[4] Carter, W. E. "The Pirquet System of Nutrition and Its Applicability to American Conditions." *Journal of the American Medical Association.* 77:1541-1546. November 12 1921

[5] See Clark, T., Sydenstricker, E., and Collins, S. D. "Indices of Nutrition." *Public Health Reports.* June 8, 1923. Reprint No. 842. Washington: Government Printing Office; "The New Baldwin-Wood Weight-Height-Age Tables as an Index of Nutrition." *Public Health Reports.* March 14, 1924. Reprint No. 907; and *Heights and Weights of New York City School Children, 14 to 16 Years of Age.* New York: Metropolitan Life Insurance Company Statistical Bureau. 1916

sion and, for at least a part of them, to do nothing to alleviate the effects of poverty on their health. Such an examination, except where records exist, is ruled out in any study of past events and the second condition would probably be impossible to carry on even if it were ever decided that such a policy should be followed strictly for the sake of the data desired.

Fortunately, however, there are indexes of the health of a population group that do not involve the assessment of the physical fitness of each of the individuals within the group. If there is a reasonably high correlation between physical fitness and the incidence of disabling illness or the extent of mortality, one can estimate the average health of a population group by the frequency of occurrence of these events and in this way secure much of the necessary information.

The limitations of such indexes of the health of population groups are set by the availability of data and the specificity and precision of the measure. In practice only three rates are used extensively—mortality, sickness, and the prevalence of physical defects including malnutrition and underweight. Each of these three measures can be expanded into many by making it specific for certain items, such as mortality from specific causes at specific ages in specified occupations or in other specific types of population groups. Under the heading of mortality there are also such specific rates as the mortality of infants under one year of age, the mortality of infants under one month of age (neo-natal mortality), and mortality of mothers as a result of childbirth, miscarriage, or abortion.

Mortality is not the most precise measure of health because it considers only that small proportion of illnesses that end fatally. Mortality data, however, are about the only figures that are available in detailed causes for specific age and sex groups over a long period of years, and they cannot be neglected in a study of health covering a series of years.

INFANT, NEO-NATAL, AND MATERNAL MORTALITY AND STILLBIRTHS

Each of these four measures of mortality is usually expressed as the number of deaths per 1,000 live births. The difficulty of obtaining accurate population estimates for intercensal years therefore does not enter into the problem since the continuous registration of births affords the necessary population base for each year or other period desired for the computation of rates. State registration offices commonly have an alphabetical file of registered births and use various methods to get complete registration. In this connection, death certificates for infants under one year of age and for mothers dying from childbirth are often checked routinely against the file of births and stillbirths to make sure that the infant's birth had been registered.

In checking infant and maternal death certificates with the birth files, data that are not on the death certificate but are on the birth certificate could be utilized in computing infant and maternal mortality rates. Among the important items of this kind are certain personal particulars about the father and mother of the dead infant and about the husband of the dead mother; namely, occupation, birthplace, color, and age. Since the certificates for births which did not result in death to the infant or to the mother carry the same information, the checking process affords data for the computation of infant and maternal mortality rates by occupation of the father, by employment status of the mother, and by nativity or racial stock of either or both parents.

With respect to stillbirths, all of these items are already available on the standard certificate of stillbirth because it contains the same information as a certificate of live birth. Thus, the same types of rates as suggested above for infant mortality could be computed for stillbirths, in so far as such rates are computed with the usual population base; namely, live births.

A. Neo-Natal and Other Infant Mortality

Infant mortality is commonly expressed as the number of deaths under one year of age per 1000 live births during the same period. This figure approximates what is termed "true infant mortality," that is, the proportion of infants born alive who die before one year of age. Neo-natal mortality is a useful subgroup, being expressed as the number of deaths under one month of age per 1,000 live births. Because about 75 per cent of the deaths under one month of age occur at less than one week of age, one might well add a rate of deaths under one week per 1,000 live births. About 60 per cent of the infant deaths occur within one month of birth.

Infant mortality has been declining at a fairly rapid rate for about 45 years.[6] In Massachusetts where records are available for many years the infant mortality rate in 1890 was 167 per 1,000 live births and in 1935 it was 48 for the same state. The rate in New Zealand has been lower but the relative decline has been about parallel to that in Massachusetts.[7]

The major causes of infant mortality are diarrhea and enteritis, pneumonia, premature birth, congenital malformations, and injury at birth. The last three causes along with congenital debility and other diseases of early infancy are responsible for roughly half of the total infant mortality; they represent chiefly the deaths that occur within the first month and largely within the first week of life. This group of causes has shown less decline than the diseases affecting the later months of the first year of life.[8] Large decreases have taken place in diarrhea and enteritis,

[6] See Whipple, G. C. and Hamblen, A. D. "The Use of Semi-Logarithmic Paper in Plotting Death Rates." *Public Health Reports.* August 18, 1922. Reprint No. 777. P. 1988; Collins, S. D. "Infant Mortality from Different Causes and at Different Ages in Nine Cities of the United States." *Public Health Reports.* February 17 1928. Reprint No. 1209. P. 395

[7] Collins, S. D. *Op. cit.*

[8] *Ibid.* Fig. 1; Dublin, L. I. and Lotka, A. J. *Length of Life.* New York: The Ronald Press Co. 1936. P. 163

tuberculosis, and the infectious diseases common to childhood.

The declining trends in infant mortality apparently continued during the depression years, at least so far as the general population is concerned. No study of trends among the unemployed or among others most severely affected by the depression has been undertaken.

The figures quoted above are all for the general population; infant mortality is considerably higher in the lower economic strata than among the well-to-do.[9] In seven cities[10] studied by the United States Children's Bureau, 1912-1916, infant mortality in families whose heads earned less than $450 per year was 167 per 1,000 live births, as compared with 59 in families whose heads earned $1,250 or more. In England and Wales in 1911, infant mortality among the children of unskilled laborers was 152 per 1,000 live births as compared with 76 for the professional, salaried, and independent class.[11]

It is highly significant that the causes of infant deaths that have declined most in the past twenty years are the ones that show the greatest variation with family income. Woodbury[12] and the Registrar General[13] both found large variation with income in gastric and intestinal diseases, in respiratory diseases, and in communicable diseases, with much less variation in diseases of early infancy.

[9] Collins, S. D. *Economic Status and Health.* Public Health Bulletin No. 165, September, 1926. Washington: Government Printing Office. 1927. Pp. 49, 50; Sydenstricker, Edgar. *Health and Environment.* A Committee on Recent Social Trends Monograph. New York: McGraw-Hill Book Co. 1933. P. 100; Woodbury, R. M. *Infant Mortality and Its Causes.* Baltimore: Williams and Wilkins Co. 1926. P. 131

[10] See Woodbury, R. M. *Causal Factors in Infant Mortality.* Washington: U. S. Children's Bureau Publication No. 142. 1925. P. 148

[11] *Annual Report of the Registrar General of Births, Deaths, and Marriages in England and Wales.* London: printed for His Majesty's Stationery Office by Darling & Son. 74th. 1911. Sections and tables on infant mortality by occupation of the father and by social classes; Collins, S. D. *Op. cit.* P. 56

[12] Woodbury, R. M. *Op. cit.* P. 148 and Collins, S. D. *Op. cit.* Pp. 52-53

[13] *Annual Report of the Registrar General of Births, Deaths, and Marriages in England and Wales.* London: His Majesty's Stationery Office. 1911. P. 88. (See note 11)

Considering rates for infants of different ages, the picture is the same—the death rate among infants over one month of age is much higher among the low income groups but the neo-natal death rate (under one month of age) is practically the same in the various income groups.[14] The findings in England and Wales are similar—the relative difference between the infant mortality in families of unskilled and of professional and independent fathers increases as the age of the infant increases.[15]

Infant mortality is generally accepted as a rather sensitive index of the sanitary status of the infant's environment. If the correlation with environmental conditions is reasonably high, we would expect to find an increased infant mortality during the depression among those elements of the population that were hardest hit by the depression, in so far as the health of the infant was not protected by sacrifices by other members of the family. Because of habits and knowledge carried over from pre-depression days which would protect the infant from a bad environment, it would not be expected that the increase in infant mortality among comfortable families which fell to poor status would be as great as the difference between poor and well-to-do families prior to the depression. However, some reflection of the reduced circumstances of the family might conceivably be expected in their infant mortality rates. To investigate and appraise the relation between the depression and infant mortality, it would be necessary to obtain rates for the specific elements of the population that were most severely affected by the depression, because in even the severest economic crisis the number of families seriously affected is not a majority of the total. Moreover, neo-natal mortality, which at present constitutes more than half of the total infant mortality, has changed relatively little in the past forty years and is relatively independent of economic

[14] Woodbury, R. M. *Op. cit.* P. 148

[15] *Annual Report of the Registrar General of Births, Deaths, and Marriages in England and Wales.* London: His Majesty's Stationery Office. 1911. P. 88; Collins, S. D. *Op. cit.* P. 57. (See notes 9 and 11, p. 14)

status; the decrease has come quite largely in the ages above one month. Any study of infant mortality should therefore consider these two age periods separately.

The study of infant mortality should be carried beyond the depression; it is possible that the viability of the infant is influenced by nutrition of the mother during pregnancy but the effect might show up only after several years of inadequate diet; thus, any change in the infant mortality rate, particularly the neonatal rate, would appear with a definite lag with respect to the peak of unemployment or other measure of economic depression.

B. Maternal Mortality

Maternal deaths are defined as those in connection with pregnancy or childbirth. In common practice these deaths refer to those during pregnancy or within six weeks to two months after childbirth, if the death shows any probable relation to childbirth. Deaths following or in connection with an abortion or miscarriage of any pregnancy duration are usually included in maternal mortality. The maternal mortality rate is commonly expressed as maternal deaths per 1,000 live births. Since many maternal deaths follow abortions, a better base would be the total pregnancies, but abortions and miscarriages are so incompletely registered that official practice has been to exclude them from the rate base. More specific rates would be provided if maternal deaths were divided into those following viable births and those following non-viable miscarriages and abortions; the latter group would include fetuses of less than approximately seven months pregnancy duration. Thus, in computing maternal mortality following viable births, only those maternal deaths that occurred after seven or more months of pregnancy would be considered. This procedure would leave a residue of maternal deaths following miscarriages and abortions with a very incomplete birth population to which they could be related, but it would give a more accurate figure for the later births.

Maternal mortality in the United States has been relatively

constant for as many years as birth records are available; there is, however, some evidence of decline since 1930. The rate is approximately six maternal deaths per 1,000 live births, somewhat over one-third of the deaths being due to puerperal infection or septicemia, and the other two-thirds to miscellaneous other causes. Neither of these causes shows any appreciable trend over the past 20 years. Comparisons with other countries must be made with caution; there is not complete uniformity in the definition of a maternal death or of a live birth.

Few data are available on maternal mortality for the different economic strata but a recent study in England[16] indicates a considerably higher rate in the depressed coal-mining areas than in the more prosperous towns of southern England. Moreover, from 1928 to 1934 there is a rise in the depressed areas as compared with a slightly declining rate in the prosperous towns. But the phenomenon of high rates in depressed areas is not universally true; some localities with much unemployment and poverty maintained a low maternal mortality throughout the period of the study. The causes of the higher mortality are ascribed by the author to malnutrition from lack of an adequate amount of proper food during pregnancy.

If the diet of the mother during pregnancy has been inadequate in quantity or essential nutrients, some increase in maternal deaths might be expected. Inadequate and unbalanced diets must have been rather widespread in the United States during the depression (see later section in this chapter) and one might expect an increase in maternal mortality, if it were possible to measure it, in that specific population group that was most severely affected. On the other hand, relief was frequent and the special consideration usually given to pregnant women in the distribution of relief may have prevented many ill effects.

In maternal as in infant mortality it is particularly important that the investigation be carried through and beyond the depres-

[16] Williams, Lady. "Malnutrition as a Cause of Maternal Mortality." *Public Health.* 50:11-19. No. 1. October 1936

sion, as the effects of inadequate diet might be cumulative and show up in high maternal mortality rates only after a considerable lag.

C. Stillbirths, Miscarriages, and Abortions

Fetal deaths are very incompletely reported, particularly those that occur early in the period of pregnancy and are commonly referred to as abortions and miscarriages. In fact, only a few states require the registration of cases in which the duration of pregnancy is less than four or five months. Even if there be included as stillbirths only fetuses that are capable of independent life, reporting is still incomplete in many areas. However, where registration is known to be reasonably complete, it might be possible to make use of the data.

In the general population the reported stillbirth rate has been relatively constant for many years at a figure of three to four per 100 live births. Surveys which include as nearly as possible all miscarriages and abortions bring this figure up to as much as 10 to 20 such dead fetuses per 100 live births[17] and recent estimates of pregnancy wastage indicate even higher percentages.[18]

Stillbirth and miscarriage rates in Baltimore in 1915, as reported by the Children's Bureau study, varied relatively little with income of the father.[19] With respect to chronology and to economic status, these rates behave like the neo-natal mortality rate, showing relatively little variation in either instance.

[17] Computed from Sydenstricker, Edgar. "Differential Fertility According to Economic Status." Hagerstown Morbidity Studies No. XI. *Public Health Reports.* August 30, 1929. Reprint No. 1312. Table 2; Collins, S. D. "Age Incidence of Specific Causes of Illness." *Public Health Reports.* October 11, 1935. Reprint No. 1710. Table 2

[18] See Stix, Regine K. "A Study of Pregnancy Wastage." Millbank Memorial Fund *Quarterly.* 13:354. No. 4. October 1935; Taussig, F. J. *Abortions, Spontaneous and Induced, Medical and Social Aspects.* St. Louis: C. V. Mosby. 1936. Pp. 364-368

[19] See Studies in Infant Mortality by the United States Children's Bureau.

D. STUDIES SUGGESTED

(1) Trend of Infant Mortality by Social-Economic Class.— The standard birth certificate carries on it the occupation of the father and of the mother if she has an occupation, or the word "housewife" if she is not gainfully employed. Occupation on the birth certificate is said to be reasonably well filled out in some states, but in others there are many omissions and in all states there are many entries that are too incomplete to classify according to a detailed occupation code. However, the specific task that the father performs and the specific industry in which he does his work is of little interest in relation to infant mortality. If the return of occupation on the certificate is definite enough to place the father in one of such general classes as professional, business, clerical, skilled laborer, unskilled laborer, or farmer, the occupation can serve as a rough index of the social-economic level of the family.

The number of births serves as a population base for neo-natal and for infant mortality rates; that is, both of these rates are commonly expressed as per 1,000 live births. With an alphabetical index to births, such as those kept in state and city health departments, it would be feasible to check for a period of 10 to 15 years the deaths of infants under one year of age against the birth file and obtain for the dead infant the occupation, color, age, and birthplace of its father and mother.

The broad occupational group would enable one to show the trend of infant mortality rates specifically for the laboring groups which were hardest hit by the depression, and to compare those trends with those for professional, business, and clerical groups that were not so frequently reduced to a relief

Rochester, Anna. *Infant Mortality, Results of a Field Study in Baltimore, Md.; Based on Births in One Year.* Children's Bureau Publication No. 119. Washington: Government Printing Office. 1923. P. 234; Collins, S. D. *Economic Status and Health.* Public Health Bulletin No. 165. September 1926. Washington: Government Printing Office. 1927. Pp. 50-51

status. One could thus determine whether infant mortality of any social-economic class showed an appreciable deviation from the expected trend during or after the depression years.

In connection with such a proposed study it should be noted that:

(a) The occupation of the father of living and of dead infants would come from the same source, namely, the birth certificate. Thus, there would be no error such as is present in occupational mortality statistics for adults where the occupation of the living commonly comes from the entries of census enumerators and of the dead from the entries on the death certificate.

(b) Although occupations change frequently, the occupation of the parents as stated on the birth certificate would refer to a period not more than 12 months prior to the death of the infant.

(c) Occupations as stated on birth and death certificates are usually too general for studies of specific hazards of industrial workers. However, for a study of infant mortality the specific occupation is not needed; the data are usable, therefore, if the occupation is sufficiently defined to throw it into the proper broad class, that is, professional, skilled, unskilled, etc. Even if skilled, semi-skilled, and unskilled labor are so poorly defined that they cannot be accurately separated, a study would probably be worth while to compare laboring groups of all types with the professional, business, and clerical group. The best procedure would be to carry a group of ill-defined laborers separate from the skilled, semi-skilled, and unskilled which could be used or discarded in the end.

(d) In so far as the birth certificate carried the occupation of the mother, it would be possible to study infant mortality by the employment status of the mother as well as by the occupation of the father.

(e) It would be possible to omit unknown and ill-defined occupations and study infant mortality in groups with known occupations without introducing any appreciable bias because both the birth and death certificates would be in the same unknown group. It would be advisable to carry along the unknown and ill-defined groups to be sure that no harm was done by using the known occupations only.

It would be of interest and worth while to consider separately in this study neo-natal (under one month) and other infant mortality (one month to one year). It might even be worth considering infant mortality under one week of age since about 60 per cent of the infant deaths occur under one month and about 75 per cent of these occur under one week of age. Data on the birth certificates would also make it possible to study infant mortality by nativity or race of the father and mother and by age of the mother and the number of prior births. It is probably important to consider some or all of these factors as many of them would be correlated with the social-economic status of the parents and should be eliminated to make a true comparison of low and high economic groups.

(2) Trend of Maternal Mortality by Social-Economic Class.— There are about six maternal deaths per 1,000 live births. The death certificate of the mother carries only her own occupation, but a check with the birth certificate of the child after whose birth she died would give the occupation of her husband.

This check would also show whether the death occurred following a live birth, stillbirth, miscarriage, or abortion, in so far as the latter classes were registered and the duration of pregnancy was stated or implied in the terms used. It would probably be advisable to consider separately the deaths following abortions and miscarriages of less than seven months duration of pregnancy and those following births of a pregnancy duration of seven months or more.

Maternal mortality is expressed as deaths per 1,000 live births; the check with the birth certificate would thus provide data for the computation of maternal mortality rates for different social-economic classes. Color, race, age and other facts about the mother would also be available from the birth and death certificates. The carrying out of such a check over a period of 10 or 15 years would give trends in maternal mortality before, during, and after the depression for different social-economic levels of the population.

(3) Trend of Stillbirths by Social-Economic Class.—A standard stillbirth certificate carries the same information as a certificate of live birth, including the occupation of the father and of the mother. Since the stillbirth rate is commonly expressed as a ratio to live births, one could compute stillbirths per 100 live births by social-economic class and thus obtain a trend of stillbirths during the depression among the laboring groups that were hardest hit by the depression, as compared with the professional, business, and clerical groups.

The definition of a reportable stillbirth varies widely in different states as does also the completeness of reports of stillbirths falling within the definition. Due to this lack of uniformity and to the incompleteness of reporting of abortions and miscarriages occurring early in the period of gestation, it would be advisable to limit the study to stillbirths of seven or more months of pregnancy. This limitation would also serve to exclude most of the illegal abortions as well as those produced for therapeutic reasons. As the duration of pregnancy is given on a stillbirth certificate, it would be possible to do this in a city or state where the certificates are fairly well filled out, assuming all reports of miscarriages and abortions to be of less than seven months of pregnancy unless otherwise stated. The state or city selected for the study should be one in which stillbirths are fairly completely reported. There are probably no localities where the reports are entirely complete, but incomplete returns might be useful in determining trends before, during, and after the depression in so far as the evidence indicated that the completeness of reporting was not influenced by the depression.

There is ample justification for considering stillbirths in relation to the depression, because of the many conditions that might influence the stillbirth rate, such as the nutrition and care of the mother during pregnancy. However, unemployment of expectant mothers who otherwise might be working away from home might be a favorable factor in the situation even though it was an enforced leisure due to lack of work.

(4) Trend of Infant Mortality by Class of District.—Many large cities now tabulate and have been tabulating vital statistics by census tracts or other small areas. Census tracts are usually too small for computing annual infant mortality rates. For many purposes, however, the smallness of the area is a distinct advantage for a census tract is usually laid out to include a fairly homogeneous population. In other words, an Italian district will usually comprise one or more separate tracts and very poor districts are generally in separate census tracts from the good residential districts. The median rental according to the 1930 census would probably give a fairly good index of the relative economic level of the different tracts. Having so rated the many census tracts within a city, those with similar median rentals could be combined into as few as three, four, or five groups. It would then be possible to compute for a series of years infant mortality rates by these groups of census tracts and thus obtain a trend for the various years prior to and during the depression for residents of the poor sections of the city as compared with those living in better sections. Green[20] has done this for Cleveland for the year 1928 and Hauser[21] has done it for Chicago

[20] Green, H. W. and Moorehouse, G. W. "Corrected Fatality Rates in Public Health Practice." *Public Health Reports.* January 24, 1930. Reprint No. 1354

[21] Hauser, P. M. Study of infant mortality in Chicago in preparation for publication. This study includes an analysis of neo-natal and other infant mortality

for a series of years. It would add considerable interest to such a study if it were carried over a sufficient number of years to obtain trends before and during the depression. New York City vital statistics are currently tabulated by health areas[22] and could be used for such a study. So far as census data are concerned 18 cities are on a census tract basis for the 1930 census and about 50 will be on such a basis for the 1940 census.[23] Some of these cities probably tabulate vital statistics by census tracts.

Data on the populations of census tracts are usually available only in census years and the character of a census tract might change considerably from 1930 to 1937. Thus, in considering trends for types of census tracts, one is not necessarily considering trends for the same population group; the depression caused many shifts from good to poor districts and much change in the type of population in a given district. However, the consideration of trends seems worthwhile in spite of shifts in the character of the population considered. So long as we deal with births as a population base, the correct numbers are currently available; the error refers only to the type of families living in the district.

In such a study it is imperative that births and infant deaths be allocated to the district of residence as otherwise births would largely cluster in districts having maternity and other hospitals. If it seemed advisable it would be possible to allocate all infant deaths to the residence of the mother at the time of the birth of the child. Although this procedure might credit a death to an area of former residence of the family, it would avoid the greater error of crediting an infant death to one area when the birth of

---

for families classified by economic status (based on census tract median rentals) for depression and predepression years.

[22] New York City monthly mimeographed vital statistics by health areas

[23] Reckless, W. C. "The Initial Experience With Census Tracts in a Southern City." *Social Forces.* October 1936

the dead child had been credited to some other area. Inasmuch as a large part of the deaths of infants occur in the first month of life, this method of allocating the infant death to its own residence at birth would seem to be the best procedure.

In this type of study, also, it would be advisable to consider neo-natal mortality separately from the mortality of infants over one month of age. Facts that are on the birth certificate, such as color, birthplace, and occupation of the parents and whether or not the mother was employed should be taken into account in judging the comparability of the various census tracts; if the data permitted, rates that were specific for some of these factors could be set up.

(5) Trend of Maternal Mortality by Class of District.—Only the largest cities would afford sufficient data to consider maternal mortality by class of district as a large population base is required to obtain reliable rates of small magnitudes.

(6) Trend of Stillbirths by Class of District.—A study by class of district could be set up for stillbirths, in the manner outlined for live births, if a city could be found where stillbirths were reported with sufficient completeness to make the study worthwhile. In this connection it should be noted that the reports on stillbirths after seven or more months of pregnancy are more complete than in the case of earlier miscarriages and abortions.

GENERAL MORTALITY, PARTICULARLY OF ADULT MALES OF THE WORKING AGES

Mortality from all causes has declined markedly during the last half century. In 1890 Massachusetts had an annual death rate of 19.4 per 1,000 population as compared with 11.4 for 1935. In the group of states that has been in the Registration Area from its beginning, the death rate has declined from 17.2 in 1900 to 11.2 in 1933. Throughout the downswing of the depression there was no interruption in this decline, the death rate for 1933 being the lowest ever recorded in this country. In 1934 and 1935 the rates were slightly higher and preliminary data for 1936 indicate a rate of about the 1930 level.[24]

Both during and prior to the depression there were fairly large annual fluctuations in the death rate. The most usual cause for a single year's rate being above the apparent normal is the occurrence of an influenza outbreak, the one type of epidemic now experienced which sweeps the country in a short time and results in a general rise in the death rate in practically all sections. These influenza epidemics are apparently independent of economic depressions and of any other known cause.

A. Mortality from Various Diseases

As might be anticipated, every cause of death does not follow the same downward trend that is shown by the general death rate from all causes. The decline in mortality has been largest among infectious diseases such as tuberculosis, typhoid, diphtheria, and scarlet fever.[25] Since these diseases take their heaviest toll among persons of the younger ages, particularly under 40 years, one finds more rapid declines in the ages under 40 years.[26] The mortality from diseases of the middle and older ages, such as those of the heart, arteries, and kidneys, and from cancer and diabetes are all increasing at a considerable rate. Conse quently one finds in the older ages little or no decline in the total death rate; in fact there is some tendency toward higher death rates at present for the oldest age groups than prevailed forty years ago.

Data like those described above could be broken down into

---

[24] "Mortality in Certain States in Recent Years." *Public Health Reports.* May 7 1937 and April 26 1935. Table 1

[25] See Sydenstricker, Edgar. *Health and Environment.* A Committee on Recent Social Trends Monograph. New York: McGraw-Hill Book Co. 1933. P. 81; Whipple, G. C. and Hamblen, A. D. "The Use of Semi-Logarithmic Paper in Plotting Death Rates." *Public Health Reports.* August 18 1922. Reprint No. 777

[26] See Sydenstricker. *Op. cit.* Pp. 151-160

still more specific groups, such as deaths from specific causes at specific ages. The breakdown that suggests itself as most pertinent would be deaths from certain causes for males of the working ages, say 20-64 years. While it would be desirable to have rates that were also specific for type of occupation, at least to the extent of a separate rate for the laboring classes, no accurate data are available for such statistics. Because of aging of the population, adjustment for age differences probably should be made unless the age groups include a span of not more than 20 years. It would be important to compile the statistics for certain states and large cities separately, and compare the trends in agricultural states with those in industrial states.[27]

Ogburn and Thomas[28] correlated an index of business activity in the United States with mortality from all causes and from certain specific causes, using a lag of various numbers of months, up to one year. The correlations were not large but were positive in practically all instances. Later, Sydenstricker[29] pointed out that the association was positive up to about 1912 but tended to be less so thereafter; he attributed the positive correlation of the earlier years to the great influx of immigrants in the years of prosperity, who for the most part were poor and lived and worked in unsanitary surroundings and consequently had high death rates. On the other hand, immigrants are usually young and young adults have a low death rate.

Monthly and annual mortality data are available in great detail for the various states and cities.[30] However, practically no

[27] Data by sex, age, state, and cause are not published in the *Annual Mortality Statistics for the United States* but the tabulated data are available in the files of the United States Census Bureau.

[28] Ogburn, W. F. and Thomas, D. S. "The Influence of the Business Cycle on Certain Social Conditions." *Journal of the American Statistical Association*. Vol. 18. N.S. No. 139. September 1922

[29] Sydenstricker, E. "The Declining Death Rate from Tuberculosis." *Transactions of the 23rd Annual Meeting*. New York: National Tuberculosis Association. Pp. 116-121. 1927

[30] See *Annual Mortality Statistics for the United States*. Washington: U. S.

data are available by occupation or any other index of economic status.

B. Mortality from Accidents

The accident death rate was less during the heart of the depression than before or after. However, a large number of unemployed persons were not exposed to the usual hazards of industrial accident and it may be that the decline in the rate is a reflection solely of this lessened exposure. Moreover, fewer commercial automobiles and trucks were on the road during the depression and street and road accidents would thus be less. A careful analysis of accidental deaths by age, sex, and type of accident ought to answer some of these questionable points.

The Metropolitan Life Insurance Company has published a detailed study[31] of mortality from accidents and other external causes among their policyholders during the years 1911 to 1930, as compared with rates in the general population of the registration area. Unfortunately the study does not include the years of the depression.

Population estimates of the general population for the years since 1930 are less accurate, so it may be necessary to await the results of the 1940 census before reliable rates can be computed. Even if that delay is necessary, it seems highly important to include the depression period of the thirties in a study of the trend of accidents.

C. Mortality from Suicide

Suicide rates increase during depressions.[32] This phenomenon has been observed for many years. The increase is said to be

Bureau of the Census. Series runs from 1900-1934; others to be published as soon as prepared.

[31] *The Mortality from External Causes, 1911-1930.* Monograph No. 3 in a 20-year mortality review. New York: Metropolitan Life Insurance Company. 1935

[32] See Dublin, L. I. and Bunzel, B. *To Be or Not to Be. A Study of Suicide.* New York: Harrison Smith and Robert Haas. 1933. P. 103

particularly great among the higher income groups where large amounts of insurance are carried, although this does not mean that there is fraud to secure insurance for the family.

The Metropolitan Life Insurance publication cited above on accidents also includes detailed data on suicide, affording an excellent predepression background for the study of suicide during depression years.

D. Studies Suggested

(1) Trend of Mortality from Various Diseases.—A preliminary examination of the trend of death rates from major causes does not suggest any very significant deviations from expectancy during the depression. However, a careful study of the course of mortality from various causes with consideration of any deviations from the long time trend that occur during or after the depression, would seem worthwhile. If the trends could be considered for specific age and sex groups or considered separately for males of the working ages it would be better. Some comparisons between agricultural and industrial states and cities with respect to trends during the depression would probably yield interesting results.

(2) Trend of Mortality from Accidents.—An examination of the trends of accident mortality before, during, and after the depression, like the one suggested above for mortality from certain diseases, would be worthwhile. It might be possible in such a study to make some estimate of the number of deaths that did not occur from industrial accidents because of the increase in unemployment and the resultant decrease in the industrial accident hazard.

(3) Trend of Mortality from Suicide.—A study of the trends of suicide rates before, during, and after the depression, like the one suggested above for mortality from certain diseases, would be worthwhile. It would be desirable to consider the rates for persons of different economic status. The amount or type of

insurance carried would be a good index of economic status and the data must be available in some insurance records if the companies themselves would undertake such a study.

(4) Occupational Mortality.—For at least five decades the Registrar General of England and Wales has issued decennially a report on mortality from various causes in different occupations.[33] These reports each cover a three-year perod centering on the census year with mortality rates from about 30 causes for each of about 100 occupations. In the more recent reports a summary of mortality in broad social-economic groups has been included. A similar report for the United States in 1930 was prepared and published by the National Tuberculosis Association.[34]

Rates for such studies of occupational mortality are obtained by relating the occupations as reported on the death certificate to the occupations as given in the census. Great care has been taken to classify the death certificate occupations in the same way as the census occupations. Data of this kind for England and Wales arc probably more accurate than for the United States because of less shifting from one occupation to another in the former countries. One is impressed, however, by the extremely

[33] See *Supplements to Annual Reports of the Registrar General of Births, Deaths and Marriages for England and Wales.* London: His Majesty's Stationery Office—

Supplement to the 85th Report. Part II. Occupational Mortality, Fertility, and Infant Mortality for the three years 1920-22.

Supplement to the 75th Report. Part IV. Mortality by Occupation for the three years 1910-12.

Supplement to the 65th Report. Part II. Mortality by Occupation for the three years 1900-02.

Supplement to the 55th Report. Part II. Mortality by Occupation for the three years 1890-92.

Britten, Rollo H. "Occupational Mortality among Males in England and Wales, 1921-23," *Public Health Reports.* June 22, 1928. Reprint No. 1233; Collins, S. D. *Economic Status and Health.* Public Health Bulletin No. 165. September 1926. Washington: Government Printing Office. 1927

[34] Whitney, J. S. *Death Rates by Occupation, Based on Data of the U. S. Census Bureau, 1930.* New York: National Tuberculosis Association. June 1934

high rate for the occupational group designated as "general laborer"; the rate is so high that it suggests the possibility that the occupations returned by the doctor or undertaker on the death certificates are less specific than those returned by the census enumerator. It is impossible to estimate the extent of any such error and if the system is transferred to the United States it will be just as impossible to estimate the error here; however, it may be presumed that it will be larger in the United States because of more changing from one occupation to another.

The above situation leads to the suggestion for a study which is not directly related to the depression but which would be in the interest of obtaining a check on the accuracy of this type of occupational mortality statistics. A fairly large city or a fair-sized state would probably be as large a job as could be done. If an alphabetical list of the census returns for 1930 for such a locality could be set up with the statement of occupation on the index card, it would be possible to check death certificates for deaths occurring after the date of the census (April 1, 1930) to this file and compare the statement of occupation given by the undertaker on the death certificate with the statement of occupation given on the census return. To make the study complete one should take also the year 1931, since three years adjacent to the census would probably be used in any occupational mortality study. The results of such a check would yield death rates by occupations but, of even greater value than that, they would give an estimate of the error in the rate that resulted from differences in the statement of occupation on the death certificate and on the census return. A study of this kind for 1930 would be an excellent preliminary step to a study of mortality statistics by occupation as based on the 1940 census. A repetition of the study to check death certificate occupations with the 1940 census would be worthwhile in view of possible improvement in occupation returns on death certificates.

It is believed that the study would not be too large to be practicable if limited to a large city or a medium sized state, and to males of the working ages, say 20-64 years.

MORBIDITY OR SICKNESS RATES

Mortality from specific causes at specific ages for each sex is available in most of the principal countries of the world. In contrast to this complete information on mortality, there are no detailed data on the extent and causes of illness for any large population group in any country.

There are several sources of morbidity records that have been used in recent years (a) special surveys of groups of families (b) records of disability among persons insured in various kinds of sick benefit associations and insurance companies (c) records of sickness among school children (d) records of hospitals and institutions (e) reports of notifiable diseases sent to local health departments and (f) special canvasses of physicians to obtain the number of cases under their care or observation on a given day, used particularly in the field of venereal disease.

A. General Population

Considering first the special family surveys, it is well to distinguish between those to obtain the prevalence of illness on a given day and those which obtain the incidence of new cases over a period of time. In the prevalence survey the housewife is asked who in her family is sick or disabled on the day of the canvass but this method disregards any case that has recovered or died prior to the day of the canvass. Thus it tends to emphasize the chronic diseases that are present on any day that the canvasser may call and to minimize the minor respiratory and other acute affections that occur so frequently throughout the year but with such short durations that few are picked up as present on the day of the visit.

This type of prevalence survey goes back to at least 1851, for such a question was a part of the Irish Censuses of 1851, 1861, and 1871, and also of the Australian Census of 1881. United States Decennial Censuses of 1880 and of 1890 carried similar questions. Tabulations were made only for selected areas in which the data seemed complete, including for 1880 twenty million persons over 15 years of age living in 19 states.[35] Later censuses did not repeat the question and the next large survey of this kind was made by the Metropolitan Life Insurance Company in the years 1915-17, covering one-half million people.[36]

Surveys of the prevalence type were made by Sydenstricker[37] in cotton mill villages in South Carolina in 1916 and 1917. In 1918, in one of these mill villages, Sydenstricker[38] made the first attempt to assemble a record of new cases of illness occurring during a given time by periodic visits to households in which was asked not only who was sick on the day of the visit but who had been sick since the preceding call of the enumerator. This incidence type of survey has been used in recent years very much more than the prevalence type, the two outstanding surveys being the Hagerstown study[39] which was made by the originator

[35] See *Tenth Census of the United States, 1880.* XII. Mortality and Vital Statistics. Part II. Sec. IX. Morbidity or Sick Rates. P. CXXXVI; *Eleventh Census of the United States, 1890.* XI. Vital and Social Statistics. Sec. XI. P. 474

[36] Stecker, M. L. *Some Recent Morbidity Data.* New York: Metropolitan Life Insurance Company. 1919

[37] Sydenstricker, E., Wheeler, G. A. and Goldberger, J. "Disabling Sickness among the Population of Seven Cotton-Mill Villages of South Carolina in Relation to Family Income," *Public Health Reports.* November 22, 1918. Reprint No. 492; Wiehl, D. G. and Sydenstricker, E. "Disabling Sickness in Cotton-Mill Communities of South Carolina in 1917" *Public Health Reports.* June 13, 1924. Reprint No. 929

[38] Sydenstricker, E. and Wiehl, D. G. "A Study of the Incidence of Disabling Sickness in a South Carolina Cotton-Mill Village in 1918," *Public Health Reports,* July 18, 1924. Reprint No. 938

[39] Sydenstricker, E. Hagerstown Morbidity Studies, 1921-24

	Public Health Reports	Reprint No.
(a) General results	Feb. 13, 1925	989
(b) The method of study and general results	Sept. 24, 1926	1113

of the method, and the study of incidence and costs of illness made by the Committee on the Costs of Medical Care[40] which,

(c) Reporting of notifiable diseases	Oct. 8, 1926	1116
(d) The extent of medical and hospital service	Jan. 14, 1927	1134
(e) The age curve of illness	June 10, 1927	1163
(f) A comparison of the incidence of illness and death	June 24, 1927	1167
(g) The illness rate among males and females	July 29, 1927	1172
(h) The causes of illness at different ages	May 4, 1928	1225
(i) The incidence of various diseases according to age	May 11, 1928	1227
(j) Sex differences in the incidence of certain diseases	May 25, 1928	1229
(k) Completeness of reporting of measles, whooping cough, and chicken pox at different ages	June 28, 1929	1294
(l) Economic status and the incidence of illness	July 26, 1929	1303
(m) Differential fertility according to economic status	Aug. 30, 1929	1312
(n) Effect of a whooping cough epidemic upon the size of the non-immune group in an urban community.	Milbank Memorial Fund *Quarterly*. Vol. 10. No. 4. Oct. 1932	

[40] Falk, I. S., Klem, M. C. and Sinai, N. *The Incidence of Illness and the Receipt and Costs of Medical Care among Representative Family Groups.* Publication No. 26 of the Committee on the Costs of Medical Care. Chicago: University of Chicago Press. 1933; Collins, S. D. Illness and Medical Care in 9,000 Families in 18 States, 1928-31

	Public Health Reports	Reprint No.
(a) Causes of illness	March 24, 1933	1563
(b) Frequency of health examinations	March 9, 1934	1618
(c) Frequency of eye refractions	June 1, 1934	1627
(d) A general view of the causes of illness and death at specific ages	Feb. 22, 1935	1673
(e) Age incidence of illness and death considered in broad disease groups	Apr. 12, 1935	1681
(f) Age incidence of specific causes of illness	Oct. 11, 1935	1710
(g) History and frequency of smallpox vaccinations and cases	Apr. 17, 1936	1740

in its morbidity aspects, was patterned after the Hagerstown study. Surveys have also been made by the Public Health Service and the Milbank Memorial Fund in a rural part of Cattaraugus County, N.Y., and in Syracuse, N.Y., but the complete results are not yet published.[41] Special and intensive studies of the incidence of the minor respiratory diseases have been made by both the United States Public Health Service[42] and the Johns Hopkins School of Hygiene.[43]

(h) History and frequency of typhoid fever immunizations and cases	July 10, 1936	1758
(i) History and frequency of diphtheria immunizations and cases	Dec. 18, 1936	1789
(j) History and frequency of scarlet fever immunizations and cases	*In Press*	

[41] Morbidity studies in Cattaraugus County and Syracuse, New York. Sydenstricker, E. and Collins, S. D. "Age Incidence of Communicable Diseases in a Rural Population." *Public Health Reports.* January 16, 1931. Reprint No. 1443; Wiehl, D. G. and Gover, M. "Epidemic of Mild Dysentery-like Disease in Cattaraugus County, N.Y., Summer of 1930." *Public Health Reports.* July 1, 1932. Reprint No. 1539; Wiehl, D. G. "Prenatal Care of Rural Mothers." Milbank Memorial Fund *Quarterly.* Vol. 9. No. 3. July 1931; Randall, M. G. "Public Health Nursing Service in Rural Families," Milbank Memorial Fund *Quarterly.* Vol. 9. No. 4. October 1931; "A Note on the Extent of Tuberculin Testing and Tuberculosis Infection in Cows in a Rural Area of Cattaraugus County," Milbank Memorial Fund *Quarterly.* Vol. 9. No. 2. April 1931; "Incidence of Contagious Abortion among Cows in Cattaraugus County," Milbank Memorial Fund *Quarterly.* Vol. 9. No. 2. April 1931; Wiehl, D. G. and Kline, E. K. "Sanitary Conditions in a Rural Area of Cattaraugus County," Milbank Memorial Fund *Quarterly.* Vol. 10. No. 2. April 1932; Sydenstricker, E. "A Study of the Fertility of Native White Women in a Rural Area of Western New York." Milbank Memorial Fund *Quarterly.* Vol. 10. No. 1. January 1932; Kiser, Clyde V. "Pitfalls in Sampling for Population Study," *Journal of the American Statistical Association.* Vol. 29. No. 187, September 1934

[42] Frost, W. H. and Gover, M. "The Incidence and Time Distribution of Common Colds in Several Groups Kept under Continuous Observation." *Public Health Reports.* September 2, 1932. Reprint No. 1545; Collins, S. D. and Gover, M. "Incidence and Clinical Symptoms of Minor Respiratory Attacks With Special Reference to Variation With Age, Sex, and Season." *Public Health Reports.* September 22, 1933. Reprint No. 1594

[43] Doull, J. A., Herman, N. B., and Gafafer, W. M. "Minor Respiratory Diseases in a Selected Adult Group: Prevalence, 1928-32, and Clinical Charac-

The data obtained by surveys of sickness conducted by periodic visits to households are limited by the extent to which the housewife recalls the occurrence of illness in her family; therefore, the frequency of the visits has a definite effect upon the number of illnesses that is recorded. In canvasses repeated at intervals of two to three months, there is usually obtained an illness rate of approximately one case per person per year. About 60 per cent of these cases are such as to render the individual unable to work, attend school, or pursue other usual activities for one day or longer, and nearly 85 per cent of these disabling illnesses are bed cases for one or more days.[44] If the contacts with the household are made as frequently as once a week and it is understood that all colds of even the mildest type are to be reported, the canvass will yield a record of about three respiratory attacks per person per year.[45] The limitation of the memory of the informant seems to be the important factor in excluding most of the mild respiratory attacks when the canvasses are made at less frequent intervals.

Obviously, the family survey method cannot be used to obtain a general record of illness for any long period prior to the survey. It is therefore inapplicable for a study of what has already happened in the recent depression of the early thirties. If one limited the study to disabling illnesses of three, four, five, or six months or longer in duration, it might be possible to obtain such a long-time record, but it would be necessary to canvass a

---

teristics as Observed in 1929-30," *American Journal of Hygiene.* May 1933; Van Volkenburgh, V. A. and Frost, W. H. "Acute Minor Respiratory Diseases Prevailing in a Group of Families Residing in Baltimore, Md., 1928-30." *American Journal of Hygiene.* January 1933; Gafafer, W. M. and Doull, J. A. "Stability of Resistance to the Common Cold." *American Journal of Hygiene.* November 1933. (See list on P. 723 for other Johns Hopkins University publications on the common cold.)

[44] See Collins, S. D. "The Incidence and Causes of Illness at Specific Ages," Milbank Memorial Fund *Quarterly.* 13:326. No. 4. October 1935

[45] See Collins, S. D. and Gover, M. *Op. cit.* P. 1158. (Note 42. P. 35)

very large number of families to obtain enough data to be of value. Moreover, long illnesses tend to break up the family by reason of the patient being sent to an institution or the illness resulting in death. Thus, a family canvass would miss illnesses that led to the breakup of the family.

For a few illnesses, such as the communicable diseases of childhood, fairly complete histories for children up to 15 or 20 years of age apparently can be obtained by asking the mother about her own children. However, these communicable diseases appear to have little relation to the depression.

The Health Inventory conducted by the United States Public Health Service in 1935-36 covered one hundred cities and was designed to obtain information, for a period of twelve months, particularly on chronic diseases of such severity as to be remembered and reported.[46] The schedule called for a record of all illness that caused as much as seven consecutive days of inability to work, attend school, or pursue other usual activities. The tabulation provides for a further subdivision of illness according to duration of disability and if there is evidence of incompleteness in the shorter cases, the final report will be concerned more largely with the cases of four-weeks duration or longer. In addition, questions were asked about a specified list of chronic diseases regardless of the duration of disability. The results of this survey will give data on cities scattered throughout the United States and will afford for the first time comparison of one geographic area with another.

The Health Inventory will also afford ample sickness data for families of different income levels but no trends or comparisons with depression years will be possible. However, the data do include rental of the home in 1929 and in 1935-36. This rental information may be useful in determining the family's economic

[46] See Perrott, G. St. J. and Holland, Dorothy F. "Chronic Disease and Gross Impairments in a Northern Industrial Community," *Journal of the American Medical Association*. Vol. 108. No. 22. May 29 1937

level in the predepression and depression year, and in setting up roughly the following groups: (a) comfortable in 1929 and in 1936, (b) comfortable in 1929 and poor in 1936, (c) poor in 1929 and in 1936, and (d) poor in 1929 and comfortable in 1936. If these groups are as useful as they appeared to be in the Health and Depression Studies[47,48] there should be, from the large sample surveyed, sufficient numbers of persons and cases to study in great detail sickness rates from a large number of specific causes. Data are available also on the duration of illness in terms of inability to work, and for chronic diseases in terms of years since the first attack. The extent of medical care by hospitals, clinics, physicians, and nurses is likewise recorded for each case.

All of the incidence studies described above as having been made since 1918 have included some index of the economic status of the family—either an estimate made by the family of the actual income during a period of time or some less precise index, such as the enumerator's opinion of the apparent economic level of the family. Analyses of the various sets of data have indicated higher rates in the lower economic levels. Not all diagnoses show the same degree of association with economic status, but the data thus far tabulated are not extensive enough definitely to set apart any disease or groups of diseases whose incidence shows significantly closer relationship to economic status than is true for the average incidence of all diseases. The communicable diseases of childhood are possible exceptions to this statement, for their relation to economic status seems to be somewhat different from other diseases.

The study of the Committee on the Costs of Medical Care[49]

---

[47] See more detailed description of these studies in a later paragraph.

[48] Perrott, G. St. J. and Collins, S. D. "Relation of Sickness to Income and Income Change in 10 Surveyed Communities." Health and Depression Studies No. 1. *Public Health Reports*. May 3 1935. Reprint No. 1684. See list p. 595 for other publications on health and the depression.

[49] Falk, Klem, and Sinai. *Op. cit.* P. 75. (See note 40, p. 34)

may be an exception to the general rule of higher sickness rates in the lower income groups. It is probable, however, that this exception is explained by the fact that the income groups considered by the Committee reached considerably further into the higher levels than in the case of the usual study. This fact, combined with the emphasis laid upon the expenditure of funds for illness, may have led to a more complete recording of minor cases in the upper income brackets. In terms of total days of disability, even this study shows higher rates in the lower income levels.[50] The Health Inventory data on these points will be of particular interest.

One study had the sole purpose of considering illness in relation to economic status and the depression—the so-called Health and Depression Studies made by the United States Public Health Service in cooperation with the Milbank Memorial Fund.[51] This survey recorded illness during a three-months' period in the early spring of 1933, but included an estimate of the annual family income in each of the four preceding years. By the use of this income history which extended back to the prosperous year of 1929, the families were classified as: (a) those which had remained comfortable throughout the period, (b) those which had fallen from comfortable to poor, and (c) those which had been poor throughout the period. The addition of a moderate group between comfortable and poor leads to more classes, but the three described above illustrate the general method used in the study. The total family income and the per capita income were both used in considering sickness.

An examination of the sickness rates according to income his-

[50] See Perrott, G. St. J. "The State of the Nation's Health," *The Annals of the American Academy of Political and Social Science.* 188:141, November 1936

[51] Perrott and Collins. *Op. cit.:* Palmer, C. E. "Height and Weight of Children of the Depression Poor." Health and Depression Studies No. 2. *Public Health Reports.* August 16, 1935. Reprint No. 1701; Wiehl, D. G. "Diets of Low-Income Families Surveyed in 1933." Health and Depression Studies No. 3. *Public Health Reports.* January 24, 1936. Reprint No. 1727

tory indicated the lowest rate among those who had remained comfortable, the next among those who were poor throughout the period, with those who had fallen from comfortable to poor having a slightly higher sickness rate. If one could assume that those who fell from comfortable in 1929 were in all respects comparable to those who were comfortable but did not go down during depression, this method would afford an excellent comparison of sickness among the depressed as compared to similar families which were not depressed. However, in an analysis of the types of families in the two groups with respect to such items as nativity, education, and occupation, it was indicated that those who shifted from comfortable to poor were, even in 1929, rather different from those who remained comfortable. Regardless of the way the phenomenon came about, it is an extremely significant finding that there was an excessive incidence of sickness in the low income depressed families who were least able to pay for medical care.

B. MEMBERS OF SICK BENEFIT ASSOCIATIONS AND OTHER INSURED PERSONS

There is a considerable mass of information on sickness among industrial workers who are members of sick benefit associations or who work for industrial concerns that keep regular sickness records for insurance or other purposes. The United States Public Health Service has collected such records from a considerable number of sick benefit associations and has issued publications based on these data for a continuous period since 1920.[52] One might think that these records would afford an

[52] Brundage, D. K. Sickness among Industrial Employees.

	Public Health Reports	Reprint No.
(a) Sickness among industrial employees, 1920-24	Jan. 22, 1926	1060
(b) Sickness among industrial employees, 1920-28	Jan. 17, 1930	1347

excellent history of sickness among the working classes during the depression, for they refer largely to industrial workers and it was the industrial population that was severely affected by the depression. Important as these records are from many points of view, they do not give a history of illness for the unemployed who represent the class that bore the brunt of the depression. Although a member of a sick benefit association may retain his membership for a few months after separation from the company, the records, in the main, are exclusive of the unemployed and particularly of the unemployed who remained so for long periods. These unemployed would be the very persons who would show a high sickness rate, partly by reason of living conditions due to lack of income but also by reason of the selective effect of the policies of industrial concerns in hiring and firing; for these policies tend to reject the unfit applicant for a job and, when depression makes it necessary to reduce personnel, to drop from the payroll those who have been frequently sick.

In spite of their limitations, these records of sickness from specific causes deserve a careful study of trends before, during, and after the depression. However, the selective tendency must always be kept in mind to avoid misinterpretation of the results.

Voluntary sickness insurance in this country must cover a very large number of individuals but those companies that carry such

---

(c) Incidence of illness among male industrial employees in 1933 as compared with earlier years — May 25, 1934

(d) Sickness among male industrial employees during the final quarter of 1935 and the entire year. — May 22, 1936 — 1748

(e) Sickness among male industrial employees during the second quarter and the first half of 1936. — Dec. 4, 1936

Brundage, D. K. "A 10-year Record of Absences from Work on Account of Sickness and Accidents," *Public Health Reports*. February 25, 1927. Reprint No. 1142; and "The Incidence of Illness among Wage Earning Adults," *Journal of Industrial Hygiene*. Vol. 12. No. 9. November 1930

insurance seldom publish the sickness records of the insured.[53] There would be, however, some selective effect in these records, in that those individuals who were most severely affected by the depression would probably have dropped out because of inability to pay premiums. Any study of such insurance records should include a careful study of the number of policyholders who dropped out during the depression. A further difficulty with such data is the fact that laborers who are the most significant class with respect to the effect of the depression on health would be poorly represented in the group of persons carrying voluntary sickness insurance.

In spite of alleged malingering, a study of cases and days of sickness among persons carrying voluntary sickness insurance seems worthwhile. The records for a five-year period of one such insurance association were analyzed by Emmet,[54] but the analysis with respect to the depression would have to consider trends over a period of years before, during, and after the depression.

C. School Children

Records of absence from school constitute a great untapped source of sickness data for the one occupation of childhood, school attendance. Every school in the United States, from the great city high school to the one-room rural school, keeps records of attendance that are surprisingly uniform in character and content, and most teachers render monthly summaries of such attendance to the school authorities.

It is true that the purpose of such records is to measure absence from all causes only, but in some school systems the cause of absence in such terms as "sick," "quarantined," and "other" are indicated. In at least the lower grades the teacher usually knows

[53] See Keffer, R. "Group Sickness and Accident Insurance," *Transactions of the Actuarial Society of America.* Vol. 28. Part I. No. 77. May 19 and 20 1927

[54] Emmet, B. "Disability among Wage Earners," *Monthly Labor Review.* November, 1919. U. S. Bureau of Labor Statistics. Washington: Government Printing Office, 1919; and "Duration of Wage Earners' Disabilities," *Monthly Labor Review.* March 1920. U. S. Bureau of Labor Statistics. Washington: Government Printing Office. 1920

not only the cause of the absence, in these general terms, but is told the specific disease by either the returning pupil or a brother, sister, or neighbor child. Some years ago one of the writers[55] outlined a method by which the school record system could be slightly modified and become the source of more or less detailed sickness data. The common communicable diseases of childhood and the minor respiratory diseases make up the great majority of the causes of sickness among school children, and the child's verbal report to the teacher or the parent's written excuse for the absence probably records these causes with a fair degree of accuracy.

The few special studies that have been made of the extent and causes of sickness among school children have usually been accomplished, not by modification of the official school record of attendance, but by setting up a more or less duplicate system which included the data desired on the causes of sickness as well as the usual record of attendance. Some early studies were made by the United States Public Health Service in various localities in Missouri.[56] These were followed by a more complete and detailed study of school absences in Hagerstown, Maryland for a period of nearly four school years.[57] The results of these various studies together with sickness data on two Florida counties are given in considerable detail in a Public Health Bulletin on *The Health of the School Child.*[58]

---

[55] Collins, S. D. "The Place of Sickness Records in the School Health Program," *Fifth Annual Meeting Transactions.* New York: American Child Health Association. 1928. P. 149

[56] Collins, S. D. "Sickness among School Children," *Public Health Reports.* July 8, 1921. Reprint No. 674; "The Relation of Physical Defects to Sickness." *Public Health Reports.* September 8, 1922. Reprint No. 782; "School Absence of Boys and Girls." *Public Health Reports.* October 27, 1922. Reprint No. 793

[57] See Collins, S. D. "Morbidity among School Children in Hagerstown, Md." *Public Health Reports.* September 19, 1924. Reprint No. 957; "Incidence of Sickness among White School Children in Hagerstown, Md." *Public Health Reports.* February 27, 1925. Reprint No. 993

[58] Collins, S. D. "The Health of the School Child." *Public Health Bulletin* No. 200. August 1931. Washington: Government Printing Office. 1931

The causes of sickness among school children can be classified into relatively few groups; thirty classes are sufficient to express them in great detail and a dozen to fifteen groups are enough for all practical purposes. The four-year record of sickness in Hagerstown indicated an average of 2.41 illnesses (absences of one-half day or longer) per child during a nine-months school term with a loss of 7.34 school days per child per school year. These illnesses had a mean duration of 3.04 school days per case and a median of 2.00 school days per case. There were 3.03 absences from causes other than sickness per child per school year, with an aggregate of 5.57 school days lost from causes other than sickness per child per school year. Absences of this kind averaged 1.84 school days each. Thus, in this Hagerstown record, 45 per cent of the periods of absence and 57 per cent of the total days lost from school were due to sickness. These figures check fairly well with the results in Missouri except for the year 1919-20, when influenza was prevalent and a higher percentage of the days lost were due to sickness.

Considered by age much more time is lost on account of sickness by the younger school children than by the older; absences from causes other than sickness increase with age to about 14 years, after which there is a decrease.

Of the total cases of illness among school children, nearly 46 per cent are due to respiratory diseases; of the total days lost on account of sickness, 43 per cent are due to respiratory diseases.

Studies made by other agencies[59] are in general agreement with the above results of studies by the United States Public Health Service.

There are no tabulated data on the trend of sickness among school children because no continuous study has been set up to

[59] See Taylor-Jones, L. "Causes of Absence in One Grade of 15 Public Schools in Washington, D.C." *Public Health Reports*. September 12, 1924. Reprint No. 954; Nesbit, O. B. "Sickness and Absence Records in the School Health Program."

give comparable records year after year. A period of three or four years, such as that covered by the Hagerstown study, is not sufficient to determine trends because there is a wide variation from year to year in the total amount of sickness. This variation is due chiefly to frequent epidemics of the minor respiratory diseases and of the common communicable diseases of childhood. Since respiratory diseases are responsible for nearly half of the total sickness of school children, such diseases loom largest in causing the annual fluctuations in school absence.

In a few instances attempts have been made to take account of the economic status of the children under consideration. In the Missouri and Florida studies mentioned above, the children were classified according to the occupation of the father into the three groups, professional and salaried and merchant class, skilled labor, and unskilled labor.[60] In the Missouri study which covered an influenza year, children of professional, salaried, and merchant fathers lost on account of sickness 5.1 per cent of the

Fifth Annual Meeting Transactions. New York: American Child Health Association. 1928; Wilson, C. C., Hiscock, I. V., Watkins, J. H. and Case, J. D. "A Study of Illness among Grade School Children." *Public Health Reports.* July 31, 1931. Reprint No. 1497; Harmon, G. E. and Whitman, G. E. "Absenteeism Because of Sickness in Certain Schools in Cleveland, 1922-23." *Public Health Reports.* June 6, 1924. Reprint No. 928; Harmon, G. E. and Whitman, G. E. "Absenteeism among White and Negro School Children in Cleveland, 1922-23." *Public Health Reports.* March 21, 1924. Reprint No. 908; Evans, C. C. "Causes of Absence in the Elementary School of the University of Chicago During Eight School Years." *Journal of Preventive Medicine.* 3:419-431. No. 6. November 1929; Mason, H. H. "Some Records Regarding Absences and Their Causes in the Lincoln School of Teachers College," *American Journal of Diseases of Children.* 22:500-507, No. 5. November 1921; Sanford, C. H. "The Causes of Absence in a Boys' School," *American Journal of Diseases of Children.* 25:297-301, No. 4. April 1923; Smith, R. M. "Respiratory Infections in School Children." *Transactions 33rd Session.* American Pediatric Society. 1921. Pp. 275-285; "A Health Study of a Boys' School." *American Journal of Diseases of Children.* 18:246-253, No. 4. October 1919

[60] Collins, S. D. *Economic Status and Health.* Washington: Government Printing Office. 1927. P. 43. Tables 17 and 18

days they were enrolled, children of skilled laborers 5.3 per cent, and of unskilled laborers 6.6 per cent. The figures for the Florida data for a non-epidemic year were 3.0 per cent, 3.7 per cent, and 4.4 per cent for children of professional and salaried people, skilled laborers, and unskilled laborers, respectively.

D. Hospitalized Illness

Hospital cases are always a selected group from several points of view: *First,* certain types of illness are more likely to get into a hospital, particularly those involving surgery. In a recent study[61] it was found that 60 per cent of the hospital cases underwent surgical treatment and, on the other hand, about 60 per cent of all the surgical cases were in a hospital. Cases without surgery, but of a similar nature, also tend to get into hospitals, for example, confinements and accident cases. *Second,* there is a selection of the more severe and complicated cases, there being a stronger trend toward hospitals for persons with two, three, and four ailments than for persons with only one disease. The considerable proportion of deaths now occurring in hospitals is largely an expression of the tendency for severe cases to be sent to a hospital. *Third,* there is an economic selection, there being more of the well-to-do in the private cases and more of the very poor in the public or free cases with the middle range of salaries underrepresented. Moreover, during depressions there is a radical shift in the proportion of cases that is private and the proportion that is free. These matters are discussed more fully in Chapter IV of this monograph.

In spite of these selective factors, the records of good hospitals afford detailed and complete illness records that should not be overlooked as a source of morbidity data. The great shortcoming that no single hospital can overcome is that there is no popula-

[61] Collins, S. D. "A General View of the Causes of Illness and Death at Specific Ages." *Public Health Reports.* February 22 1935. Reprint 1673. P. 244

tion that can be used in computing morbidity rates. Because of this shortcoming, an increase in hospital cases may mean the growth of the hospital rather than an increase in morbidity. Nevertheless, there are other rates that are of interest that can be computed for any hospital, such as: (1) fatality rates for specific types of cases, (2) proportionate morbidity rates expressed as the percentage of all cases that are due to certain specific causes. If these rates are made specific for sex and age, they will yield interesting results from the point of view of morbidity.

One method of overcoming the lack of a population would be to obtain uniform morbidity statistics for all the hospitals in a city or state. Such a plan was recently discussed by Bolduan.[62] If the nonresident cases were excluded and the city were large enough to hospitalize most of its own sick, one could then compute rates with the city's population as the base. Mental hospitals for some states have a unified system of hospital statistics and can thus compute rates with the state's population as the base. Such a system applied to all other hospitals would afford data on hospitalized cases of specific kinds and the trends of such rates over a period of years.

E. Reportable Diseases

The diseases that are reportable by law to the health department are mostly communicable.[63] The reports are generally recognized to be very incomplete because of several factors: (a) many cases are not attended by physicians and it is the physician who is responsible for making the report, (b) the physician's attention is concentrated upon the treatment and cure of the case and other considerations may be overlooked, (c) the patient

[62] Bolduan, C. F. "Compiling Morbidity Statistics," *The Modern Hospital.* 45, No. 2. August 1935

[63] Fowler, E. "Laws and Regulations Relating to Morbidity Reporting." *Public Health Reports.* Supplement No. 100. 1933

himself may urge the physician not to report the case.

Apart from these usual causes of incompleteness, it is probable that reporting in the midst of an epidemic is better than in non-epidemic periods when cases occur only sporadically.

Cases actually reported are summarized by the state health departments and the summaries forwarded to the United States Public Health Service for publication.[64] In spite of incompleteness some good studies have been made from these reports supplemented by mortality and other records.[65] Some examination of trends in different states and cities might be worthwhile.

F. Venereal Diseases

Because of the extreme incompleteness in venereal disease reporting, a method has been devised of securing for a given day a more complete total of all cases under treatment or observation by physicians, clinics, hospitals and other healing institutions.[66] Complete returns from all doctors in a city or state make it possible to compute rates based on the total population, but non-resident cases being treated within the city are a source of error. Since the origin of this method, such surveys have been made in many places by the United States Public Health Service and others[67] and have been summarized by Parran and Usilton.[68]

[64] *Annual Summaries of Notifiable Diseases* published as reprints from or supplements to the *Public Health Reports* (a) in states (b) in large cities (c) in small cities

[65] See Hedrich, A. W. "Changes in the Incidence and Fatality of Smallpox in Recent Decades." *Public Health Reports.* April 3, 1936. Reprint No. 1738; "The Movements of Epidemic Meningitis 1915-1930." *Public Health Reports.* November 13, 1931; Sydenstricker, E. "The Rise in the Meningococcus Rates, 1926-28." *Public Health Reports.* September 28, 1928

[66] Brunet, W. M. and Edwards, M. S. "A Survey of Venereal Disease Prevalence in Detroit," *Venereal Disease Information.* VIII. Pp. 197-208. No. 6. June 20 1927

[67] See Pfeiffer, A. and Cummings, H. W. "A Second Study of the Prevalence of Syphilis and Gonorrhea in Upstate New York," *Venereal Disease Information.* Vol. 12. No. 11. November 20 1931. Reprint No. 39

[68] Parran, T. and Usilton, L. J. "The Extent of the Problem of Gonorrhea and

The original idea was to resurvey the same cities and determine trends, but the results to date have been rather inconclusive[69] so far as trends are concerned.

The requests made upon doctors, clinics and hospitals on these surveys included little except the number of cases; in all instances these cases were to be classified by stage of the disease for both gonorrhea and syphilis. In the South, a color classification was also asked because of the great difference in the rates for white and Negro. In the most recent surveys more information was obtained about cases by listing them individually on a schedule and giving for each one such items as age, sex, color, occupation, and whether or not a resident of the city. Venereal disease is so largely in the hands of clinics and specialists that the average doctor can easily supply this information for his few cases and the specialists and clinics can be given clerical or other assistance by the survey in tabulating and classifying their cases in this manner. If such a method proves to be generally applicable, much additional and valuable information can be secured on venereal diseases.

Surveys of this type can be successfully carried out only by an organization that has the complete confidence of the medical profession.

G. Studies Suggested

(1) Sickness among Families Unemployed During the Depression but Now Re-employed.—It has been seen that the limitations of the sickness survey in the general population make it inappropriate for the study of sickness during a depression the depth of which is now past. Therefore, little is proposed in the way of studies involving house-to-house canvassing at the

Syphilis in the United States." *Journal of Social Hygiene.* Vol. 16. No. 1. January 1930

[69] Usilton, L. J. "Trend of Syphilis and Gonorrhea in the United States Based on Treated Cases," *Venereal Disease Information.* Vol. 16. No. 5. May 1935. Reprint No. 51. P. 161

present time. However, one extension of the Health and Depression Study suggests itself.

The Health and Depression Study[70] was made at about the most severe period of the depression. The income classes of families did not include any group that had been unemployed and poor but had been reemployed and were again on a comfortable basis; this group was not considered for the very good reason that few of such families were found in 1933. It is anticipated that, at present, a large number of families of this category could be found and a repetition of the Health and Depression Study would seem worth while with the addition of this reemployed group. The sickness rates for such a group of reemployed persons would be worth studying in comparison with rates for those who have had jobs and have been comfortable throughout the depression, and for those who were comfortable before the depression but became unemployed and are still unemployed.

(2) Trend of Sickness among Insured Persons.—In spite of selective factors that eliminate the unemployed and depressed from insurance and sick benefit schemes during the depression, it seems worthwhile to make a rather intensive and detailed study of trends of sickness rates from specific causes before, during, and after the depression. So far as data on sick benefit associations have been assembled by the United States Public Health Service[71] such a study would not entail any great amount of work. However, it would seem highly desirable to supplement this by records of insurance companies who are in the business of supplying sickness insurance and whose records would afford a much larger body of data. Insurance companies

---

[70] Perrott, G. St. J. and Collins, S. D. "Relation of Sickness to Income and Income Change in 10 Surveyed Communities." Health and Depression Studies No. 1. *Public Health Reports.* May 3 1935. Reprint No. 1684. See list p. 595 for other publications on health and the depression.

[71] See Brundage, D. K. Studies on Sickness among Industrial Employees. (See publications of Brundage listed in note 52, p. 40)

would probably refuse access to their records to outside agencies, so the analysis would have to be made by either individual insurance corporations or in some cooperative study by agencies such as the Association of Presidents of the Life Insurance Companies and the Actuarial Society of America, which cooperated in the analysis of mortality records from a large group of companies. Aside from the light that these studies might throw upon health during the depression, they would afford a body of much needed data on sickness in the United States. If analyzed with respect to age, sex, occupation, and community of residence, it would be possible by an adjustment process to estimate sickness rates for the general population in spite of the unrepresentative age and occupation distribution of the insured population. Moreover, if rates are computed for a series of years and the analysis is carried out by occupational class, it would be possible to compare trends of sickness for the laboring groups with those for the professional and salaried class.

The United States Public Health Service is now collecting a large body of data for sick benefit associations which would supplement the results of a study of insurance companies' sickness records. The records, however, cover only the period 1930-34 and would not supply the need for a trend of sickness before, during, and after the depression.

(3) Trend of Sickness among School Children.—Simple changes in the methods of keeping school attendance records could provide a large mass of statistical data on sickness among school children. Even without those changes it is believed that much valuable data could be tabulated from school records. The rates that could be obtained by a tabulation from the school records as now kept in some cities are:

(a) number of absences per 1,000 children per school year separated into those from (1) sickness and (2) other causes;

(b) days lost from school per 1,000 children per school year separated into those due to (1) sickness and (2) other causes; and

(c) percentage of absences and of days lost that are due to (1) sickness and (2) other causes.

Some school records do not distinguish between sickness and other causes of absence but many of them make such a distinction. One pitfall that may be encountered is a practice followed in some states of striking a child's name from the school roll after a given number of days of absence and re-entering him as a new pupil if he returns to school. Such a temporary drop from the roll, however, usually shows why the child was dropped, so absence from sickness and from other causes during this period can be added to the record to make it include not only "official" absence but these temporary "discharges" also.

In some school systems absences of three or more consecutive days are investigated by truant officers and their reports may include the disease causing the absence.[72] In such a city it would be possible to obtain trends of sickness from specific causes if the accumulated records could be tabulated for a series of years.

Because every teacher records annually the occupation of the father of each child, it would be possible to compute sickness rates for different social-economic groups of the population and thus secure trends of sickness before, during, and after the depression for children of laborers as compared with children of professional and business people. These school records would, therefore, provide the one source of sickness records of an unselected population that could be carried in a uniform way over a series of 10 or 15 years. Business firms exercise a selective effect on their employee populations by personnel policies regarding the type of individual to be hired or fired or dropped when depression comes. But for school children these selective factors are not operative and aside from lack of clothing to wear

[72] Wilson, Hiscock, Watkins and Case. "A Study of Illness among Grade School Children." *Op. cit.* (Note 59, p. 45)

one would not expect the depression to have any selective effect on the type of children who attended the elementary schools.

Because schools are located all over a city and serve areas of varying economic status, it would be possible for some one familiar with a city to classify the schools according to the average economic level of the district served by them. The median rental in census tracts might be useful in this connection in spite of the fact that school districts and census tracts are not coterminous. With an economic rating for each school district, one could combine the schools of a city into a few groups and consider sickness and the trend of sickness before, during, and after the depression in poor as compared with well-to-do sections of the city.

Because sickness varies with age and because of other differences, grade and high schools should be considered separately in any study of school absences. For similar reasons, the lower and upper grades of the grammar school should be separated as a substitute for tabulations by age. The days lost per case of illness is larger for younger than for older children, but in considering the trend over a period of years the whole grammar school could be put together provided there was no radical change in age distribution of the children, such as might be caused by the addition of kindergarten or the changing of the seventh and eighth grades from grammar to junior high school. The best plan would be to include the same grades for the whole period studied.

Inasmuch as the official school records are in the hands of the educational authorities and probably could not be obtained by other agencies, any regular and continuing setup to tabulate these records would probably have to be under the direction of, or at least in cooperation with, the school authorities. If the school health work is under the direction of the health department, cooperation between the school and health authorities in the study of school absences would seem feasible.

(4) Trend of Hospitalized Illness.—Even without an economic classification of the cases, it would be of interest to have the trend of specific kinds of hospital cases before, during, and after the depression. If in any city all or nearly all hospitals could be brought into such a scheme, a study of their records to obtain such rates would be worthwhile. Probably the trends of specific causes of illness would be too laborious a task to carry back of the depression, but it might be possible to consolidate hospital reports and get a trend of hospital cases from all causes, and a trend of clinic cases of all kinds. The trend in the total hospital days and the total of clinic visits before, during, and after the depression would also be worth computing if available records could be found. Such rates probably would reflect changes in hospitalization practice more than the incidence of morbidity; therefore the subject is considered in more detail in Chapter IV.

A study of this kind probably would be impossible in most places because of the necessity of getting the cooperation of all of the hospitals and clinics in the city.

(5) Case Fatality of Hospitalized Cases.—Every hospital has some index of the economic status of its patients—occupation, price paid for room, etc. The same doctors supply the ward service as have private cases elsewhere in the hospital. If the medical service can be said to be comparable, it would be of interest to compare the case fatality of cases of high and low economic status. Care would have to be exercised to see that the comparison related to comparable cases with respect to stage on entrance, age of patient, etc., and in other ways to control entrance factors. If undernutrition and other deleterious conditions undermine the resistance of the poor, one would expect them to show a higher case fatality in similar cases similarly treated. Such a study might be done in a few hospitals with exceptionally good records.

(6) Trends in Hospitalization of Mental Cases.—Although

hospitalization of mental cases depends to some extent on availability of hospital facilities, a study of trends of such cases before, during, and after the depression would seem worthwhile. Data of this kind are available annually for some states and are available annually since 1926 in biennial United States censuses of mental patients in state hospitals.[73]

A number of cautions must be exercised in the conduct of such studies. First it is necessary to insure the comparability of diagnoses in the different mental hospital units. Diagnostic variations within a state or even within a city at a given moment, or over a period of time, may greatly limit the value of combined data. It is possible also that budget and space pressures resulted in changed admission standards during the depression, with respect to marginal cases, particularly of the senile type. It may be that a time series based on admissions to psychopathic receiving hospitals would result in more nearly comparable data than time series based on data from hospitals for the permanent care of mental cases.

(7) Extent of Non-Hospitalized Mental Cases among the Depressed and Poor as Compared with the Well-to-do.—Incomplete as are the data on morbidity, they afford a much better picture of physical illness than is available for mental illness. In a few communities surveys are under way or have recently been completed to find the extent of the mental disease problem. Such a survey was made in a portion of Baltimore (Eastern Health District) and one is in progress in a Kentucky city and its county. Further surveys to obtain such basic information on the magnitude of the mental problem are important.

(8) Social Effects of Chronic Disease.—The various studies of sickness that have been mentioned all include both acute and chronic ailments. The proportion of the total cases that is due to chronic diseases varies with the type of survey. With the ex-

[73] *Census of Patients in Hospitals for Mental Disease.* Publications of the U. S. Bureau of the Census

ception of the surveys in Massachusetts[74] and New York City[75] and the National Health Inventory[76] few studies have emphasized the chronic diseases. Since the questions necessary to obtain this information and the tabulating procedures for analyzing the data are definitely different for chronic and acute ailments, adequate and complete data on chronic diseases can be obtained only by surveys directed particularly toward the chronic diseases.

The recording of the prevalence and severity of the various types of chronic diseases is in itself well worthwhile. From a sociological standpoint, however, it may be of even greater interest to determine the social effects of these chronic ailments, such as the ability of the affected families to remain self-supporting and the effects upon such matters as the education of the children of individuals who become disabled by chronic diseases. Of special interest, also, is the question of the breakup of families when the housewife or household head enters a hospital for extended treatment. In so far as data could be obtained for several years back, it would be of interest to relate these items to the depression. The depression plus the usual prevalence of chronic disease probably would have greater effects than in the case of chronic diseases that occur in normal times.

(9) Abortions and Miscarriages.—The securing of a complete history of all births, stillbirths, miscarriages, and abortions experienced by a woman is not an easy matter. Undoubtedly such a record can be obtained only by superior canvassers who can secure the confidence of the individual being interviewed. Experience has indicated that the only way to get an approxi-

---

[74] Bigelow, G. H. and Lombard, H. L. *Cancer and Other Chronic Diseases in Massachusetts.* Boston and New York: Houghton Mifflin Co. 1933

[75] Jarrett, M. C. *Chronic Illness in New York City.* I and II. New York: Columbia University Press. 1933

[76] Perrott, G. St. J. and Holland, Dorothy F. "Chronic Disease and Gross Impairments in a Northern Industrial Community." *Journal of the American Medical Association.* Vol. 108. No. 22. May 29 1937

mately complete record of abortions and miscarriages is to obtain also a complete record of live and stillbirths, and to question especially any periods between live births that are longer than the average interval. In the Cattaraugus County survey, reasonably complete information was obtained by this method.[77] If along with this information, one could obtain a rough approximation of the income level or general economic status of the family over a period of five or six years before and during the depression, it might be possible to make some estimate of the changes in the frequency of abortions and stillbirths that occurred during the depression.

ACCIDENTS

The United States Bureau of Labor Statistics has for many years collected accident data for various industries. In their *Handbook of Labor Statistics* there is a discussion of the sources of accident data.[78] These sources include reports of State Industrial Accident Boards and Commissions and other organizations administering state compensation laws; many federal agencies collecting accident statistics; and reports of the National Safety Council.[79] Summaries of accident rates in many industries are also included. At the back of the Handbook is a list of preceding publications of the Bureau with those on accidents assembled in a single series.

According to *Accident Facts,* published annually by the National Safety Council, there were 9,340,000 non-fatal disabling injuries and about 100,000 fatal injuries in the United States in 1935, involving a wage loss of $1,840,000,000; medical ex-

---

[77] See references listed in note 41, p. 35; see also Taussig, Frederick J. *Abortion, Spontaneous and Induced, Medical and Social Aspects.* St. Louis: C. V. Mosby. 1936. Chapter XXIII

[78] *Handbook of Labor Statistics.* Washington: U. S. Bureau of Labor Statistics Bulletin No. 616. 1936. Pp. 273-314

[79] *Ibid.* Pp. 273-275

pense of $360,000,000; and an overhead insurance cost of $190,000,000.[80]

More than a third of the fatal injuries were motor vehicle accidents. Automobile accidents per 1,000 population have increased greatly in the past 30 years as the number of cars increased. Other accidents, however, have been decreasing, presumably as the result of safety work which followed the introduction of the workmen's compensation system. Whether these trends were maintained at the same general rate during depression years is a matter that needs study. So far as fatal injuries in the general population are concerned, the rates for depression years were exceptionally low, but the lessened exposure to industrial hazards by the unemployed may have been the cause of the decline rather than a real decrease in the accident hazard.

If gross accident rates are computed for fatal or for non-fatal cases during the depression years by using as a base the total population or the total number of employees, it is almost certain that the rates will be less for depression years than for years before or after the depression. The decreased rates for depression years may be attributed to the depression but the result comes about from the circumstance that fewer people are working in industry and those at work on a part-time basis are putting in fewer hours of exposure to accidents.

To put accident rates of different industries on a comparable basis and avoid the biases mentioned above, industrial accident statistics have for many years been expressed as rates per million hours of exposure rather than as per 1,000 employees or population.[81] This refinement is particularly necessary in considering trends of accidents before, during, and after the de-

---

[80] *Accident Facts.* 1936 Edition. Chicago: The National Safety Council. 1936. (Data are for the year 1935.) Pp. 4, 54

[81] *Standardization of Industrial Accident Statistics.* Washington: U. S. Bureau of Labor Statistics Bulletin No. 276. 1920

pression because of the large number of part-time workers during the depression which would tend to reduce the accident rate for depression years if it is expressed as per 1,000 employees.

The Ohio State Department of Industrial Relations, Division of Labor Statistics, receives and publishes, for all firms employing three or more persons, reports showing the number of employees, the hourly wage rate, and the total wages paid to employees, classified in broad occupation groups. From these figures it seems feasible to approximate the total hours worked to serve as a base for computing accident rates from accident reports for the same establishments. These figures are available for a considerable period before as well as during the depression and are worth considering as sources of accident data. Reports on mines and quarries in Ohio give the total hours worked and thus furnish directly the necessary facts for computing accident rates per million hours worked.

Accident rates and trends in rates for different industries would be of great interest in comparison with the studies of depression changes in plant safety and hygiene, hours of work, and rates of production which are suggested in Chapter III.

The assembly and thorough analysis of all possible accident data for depression years as compared with similar rates for years before and after the depression would be of value.

NUTRITION, HEIGHT, AND WEIGHT

The health of the population is dependent in many ways upon the nature of the physical environment in which the population lives and works. The availability of food in adequate quantity and variety is of great importance because of the significance of nutrition in the well-being of the human organism. Malnutrition is not a disease, but like disease it affects the organism and may lead to varied and serious effects among those who are insufficiently or improperly nourished. Immediate results may be felt in loss of weight or in failure to gain weight, in fatigability, in changes in muscle tone and in other ways that render victims less able to pursue their normal activities. While the relationship between nutrition and disease is far from clear, it is universally recognized that rickets, pellagra, and scurvy are due to dietary deficiencies. There is ample evidence also that undernutrition weakens resistance to tuberculosis and to many other chronic and acute affections; and that it is detrimental to the patient in recovery from surgical operations and certain heart ailments. Hence, marked degrees of malnutrition among children may be significant for the reason that these future generations of workers are not developing the immunity of normal healthy persons to disease.

While it may be supposed that large scale restriction of income would result in much undernutrition, attempts during the depression to counteract its results may have been effective. Food may be the last item of the family budget to be cut down in times of reduced income. Under pressure, housewives may become more efficient in the use of food and may make better selections. And, perhaps, being aware of the danger of undernourishment, the community, through its relief organizations, health departments, and school boards, may have greatly extended its efforts to assist housewives in securing and preparing proper diets. Of this fact there is some evidence in that diets of families on direct relief appeared to be slightly better than those of others in the poor group.[82]

Whatever may have been the effect of the depression on the nutrition of the population, it is a field of research that merits attention. Unfortunately, it is an extremely difficult field in which to determine the true situation. Nutrition is difficult to

[82] See Wiehl, D. G. "Diets of Urban Families with Low Incomes." Milbank Memorial Fund *Quarterly.* 12:347, 355. No. 4. October 1934; "Diets of Low Income Families in New York City," Milbank Memorial Fund *Quarterly.* 11:314, 317, No. 4. October 1933

define. There are many kinds of nutrition, some or all of which may be affected by the quality and quantity of food intake. For these and other reasons, the nutritional status of a given organism or the average nutritional status for a population group is difficult to measure. Virtually all of the present techniques are open to criticism, as will be seen later. And finally, nutritional status is related in many subtle ways to disease and the ability of the organism to maintain health, and few of these relationships have been completely investigated. Some of the effects of malnutrition such as predisposition to disease, do not always appear immediately and hence are not practical subjects for research at the present time.

A. Underweight

The most obvious, the easiest, and the cheapest method of assembling gross evidence of undernourishment is that of weighing persons of the same age and height. Dr. Carroll Palmer of the United States Public Health Service used this technique in several recent studies in Hagerstown, Maryland and elsewhere. In the first one he compared the weights of white children in the first six grades of the public schools in 1933 with the average weights of the white children who were in the same grades between 1921 and 1927.[83] Next he compared the gain for one year in weight of children observed in both 1933 and 1934 with similar data for a predepression period, differentiating his subjects according to economic status of the parents.[84] The third study set up a comparison between the weights of children in families with practically full employment, families with no employment, and families receiving aid.[85] Finally, the weight

[83] Palmer, C. E. "Growth and the Economic Depression." *Public Health Reports.* October, 1933. Reprint No. 1599

[84] Palmer, C. E. "Further Studies on Growth and the Economic Depression." *Public Health Reports.* December 7, 1934. Reprint No. 1660

[85] *Ibid.*

increments from 1929 to 1933 of children between 6 and 14 years of age in families that had remained economically comfortable during the entire period were compared with increments of children in families that had suffered a decline in economic status and with the gains in weight of children in families that were at a low economic level throughout the period.[86]

These data for all children without differentiation on the basis of economic status revealed that between the depression and the predepression (1921-27) groups neither average weight, nor variability of weight, nor proportion of children below average weight, nor weight increment was significantly different among boys and that they were only slightly different among girls. The depression year 1933-34 was a poor growing year but there were individual years in the predepression period studied when growth was just as small, as a result of which Palmer concluded that the deficiencies of 1933-34 could not be attributed to the depression.

When subdivision according to family economic status was introduced, one set of data revealed no greater differences in weight or in gains in weight among the economic classes in depression years than appeared to exist in nondepression years. Another set of data, however, indicated that children in families with no employment weighed less than children in families with one parent employed, and that children in families receiving aid weighed less than children in families that were economically independent. And a third collection of statistics, differentiated on the basis of economic status, seemed to show that children in families whose income had fallen to a low level between 1929 and 1933 gained less weight over that period than did children in families where the economic condition, whether comfortable or poor, did not change.

Although the results of these studies are inconclusive, they

[86] Palmer, C. E. "Height and Weight of Children of the Depression Poor." *Op. cit.* (See note 51, p. 39)

point to a number of factors that must be given thought in connection with methods used and the interpretation of this kind of data. First, perhaps, is the importance of differentiation according to economic level. The failure to find evidence of nutrition changes when the comparisons were observed without this differentiation is not proof that changes did not occur among particular groups of the population studied. This seems to be demonstrated by the last mentioned set of data which showed that among families whose income had fallen to a low level the changed condition was reflected in the small weight increments of their children. And there is logical justification for this finding. Social workers state that families which were once in comfortable circumstances are notably reluctant to avail themselves of public financial assistance, whereas chronically low-income families have applied for and received relief without hesitation. It is possible that abnormal gains in weight among children of assisted families completely offset the reduced increments characteristic in the families of the new poor. Palmer recognized this problem and introduced appropriate economic classifications whenever possible.

This consideration points to the need for control of still another factor, namely, the variation among political subdivisions in standards and kinds of relief. There is ample evidence that in some places extraordinary efforts were made to counteract the danger to health of reduced incomes while, for various reasons, the efforts made in other places were much less potent. Palmer stated, in explanation of the failure to find evidence of serious nutritional problems in the Hagerstown data, that the distribution of school lunches and other forms of relief appeared to have been exceptionally well handled in that city. Because of the variation in relief policies, generalizations based on any single locality are dangerous.

A third problem arising in connection with the procedure under discussion is the inadequacy of weight (with only height and age controlled) as an index of nutritional level. Weight is a function of body build and of living habits. Moreover, not all deficiencies are revealed by weight; an animal organism may receive the necessary number of calories but nevertheless be malnourished because of deficiencies in vitamins, minerals, etc. For both of these reasons, the depression may have brought many subtle or borderline cases of malnutrition that are not discoverable by means of a technique as grossly defined as the present one. Observance of body build is particularly essential when the population under observation is not racially homogeneous. Since Palmer studied only white children in a small southern community, his data were probably not distorted by this factor; weights of children in an industrial city of the North would almost certainly be influenced by this factor.

B. NUTRITION AS JUDGED BY CLINICAL MANIFESTATIONS

Studies of undernourishment based upon even superficial medical examinations overcome the difficulty of dependence upon weight alone. Physical examinations in nutrition studies ordinarily give attention to weight, estimates of the amount of subcutaneous fat, quality of muscle tone, fatigability, and the texture and color of the skin. State and local health departments and school authorities have done most of the work in which this procedure has been employed, and Dr. Martha Eliot has summarized several of these studies.[87] Examinations of school children in New York City, Detroit, Cleveland, Fall River, fourth class school districts in Pennsylvania and elsewhere indicated almost uniformly that between the years just prior to 1930 and the period from 1931 to 1933 there were striking increases in the incidence of undernourishment among the members of this population group. Description of the method of conducting these projects is not available but two other investigations have

[87] Eliot, M. M. "Some Effects of the Depression on the Nutrition of Children." *Hospital Social Service*. 38:585-598. No. 6. December 1933

been reported in some detail. Between 1928 and 1932 the Diagnostic Clinic of the Community Health Center in Philadelphia recorded instances of malnutrition among patients under 20 years of age. Esther Jacobs[88] reported that between 1928 and 1932, patients with faulty nutrition increased from 23 to 36 per cent of all patients and that the increases were much greater in the younger age groups.

Physical examinations for evidence of malnutrition among a small group of school children in New York (Manhattan) were a part of the Health and Depression Studies of the United States Public Health Service and the Milbank Memorial Fund. Doctors Stix and Kiser reported upon this phase of the study.[89] Their conclusions were that malnourished children were found more frequently in relief families and in low income non-relief families than in higher income groups; that children in home relief (commodity) families were slightly better off than those in work relief (cash) or in non-relief low income groups. Likewise, families with consistently low incomes and families that suffered income losses during the depression had higher proportions of poorly nourished children than did families whose incomes increased or remained at a high level during the depression.

While physical examinations are superior to weight studies in that they cover a greater variety of indexes of poor nutrition, they have the limitation of subjectivity. Current techniques provide no accurate method for objective measurements of the amount of fat, of muscle tone quality, of the extent of fatigability or of changes in the color and texture of the skin. Despite all efforts to establish standards of measurement, the estimates of examining physicians are necessarily based on individual

[88] Jacobs, Esther. "Is Malnutrition Increasing?" *American Journal of Public Health*. P. 786. August 1933

[89] Kiser, C. V. and Stix, R. K. "Nutrition and the Depression," Milbank Memorial Fund *Quarterly*. 11:299-307. No. 4. October 1933

experience and judgment. This becomes particularly serious when more than one examiner is employed or when the composition of the examining group changes as it did during the course of the project reported by Jacobs, and as it undoubtedly did during the periods covered by the data of the school and health departments.

C. Derived Effects of the Depression

Thus far, the studies seen by the writers and those reviewed in the preceding section have been undertaken for the purpose of determining whether the incidence of undernourishment has increased as a result of restricted purchasing power during the depression or whether there has been no change or an actual decrease in incidence by reason of extraordinary efforts on the part of relief agencies. No researches were discovered which were designed to carry the problem to the point of revealing the existence of derivative effects of an increase in the amount of undernourishment. This appears strange, for undernutrition assumes importance because it is a precursor of disease, because it retards recovery from disease, and because it interferes with the normal activities of life.

Perhaps there are several explanations of the apparent failure to undertake research along these lines. Investigators may feel that, until they have determined that there has been an actual increase in the incidence of malnutrition, they would rather not enter the more subtle field of estimating the effects of nutrition changes themselves. Perhaps, as in the case of disease and recovery from disease, it is too early to expect the effects of malnutrition to appear. Students of school problems may have noticed increases in retardation and absences but may not have linked them with nutrition deficiencies.

D. Studies Suggested

The importance of adequate nutrition and its vulnerability to depression influence make it a promising research field. If studies could be conducted along parallel lines over a wide area, the conclusions would be available for populations living under a variety of circumstances. The depression has affected various social, economic, and geographic segments of the population in different ways and in different degrees. Relief policies and other counteractive efforts have varied enormously in kind and in intensity. The effect of these variations is that no study restricted to a small population in a localized area can have general applicability.

The need for a uniform plan was expressed by the Health Organization of the League of Nations in a report published following a conference of pediatricians and nutritionists in 1932.[90] This conference divided methods of measuring degrees of malnutrition into three groups classified according to the complexity of technique involved and the degree of refinement needed. The first group consists of body weight in relation to age, height, and other external measurements. These measures can be applied wherever subjects are found but are limited in the ways cited earlier in the present paper. The second group of determinants include estimates of the amount of subcutaneous fat, color and texture of skin, hair texture, fatigability, and muscle tone. These examinations can also be conducted where the subjects are found but they are subjective. Finally, the conference suggested a number of laboratory tests for specific kinds of deficiency such as anemia, rickets, vitamin A, and disturbance of water metabolism. These tests which include urinalysis, observation for rapid increase in weight on hospital diet, and examination for sub-periosteal hemorrhages, can be made only when it is possible to hospitalize the subjects. It is obviously

[90] *Interim Report of the Mixed Committee on the Problem of Nutrition.* Geneva: League of Nations. 1936. Vol. I, ch. IV; and report of the conference at Berlin, December 5 to 7, 1932, on the most suitable methods of detecting malnutrition due to the economic depression. *Quarterly Bulletin of the Health Organization.* Geneva: League of Nations. Pp. 116-129, March 1932

important that some method such as the first two suggested here be standardized for general use in the hands of different groups of investigators working independently of one another.

It seems appropriate to report at this time that experiments of this kind are already under way. Dr. Eliot, of the United States Children's Bureau, has collected records of weight, physical measurements, and physical examinations of a group of New Haven children and is conducting rigorous tests to determine the extent of correlation among several nutrition indexes. Dr. Palmer of the United States Public Health Service obtained physical examinations and the measurements for applying the ACH[91] index of nutrition to a group of Hagerstown school children in the Spring and again in the Fall of 1935 for the purpose of determining the extent of agreement between the two types of measures. It appears important in connection with efforts to measure the effect of depression on nutrition that these attempts to arrive at actual measures of the state of nutrition should be encouraged to go far beyond their present development.

It was stated in the preceding section that measures of general nutritional status are likely to overlook undernourishment affecting particular parts of the organism, because some deficiencies cannot be observed by weight records or even by medical examinations of external conditions. It can also be reported that work is proceeding in the direction of measuring specific types of undernourishment and their effects. Jeans and Zentmire[92] at the University of Iowa have been developing tests for discovering

[91] The ACH index of nutritional status depends upon the arm, chest, and hip measurements. It was devised by the American Child Health Association. See Franzen, R., and Palmer, G. T. *The ACH Index of Nutritional Status.* New York: American Child Health Association. 1934

[92] Jeans, Philip C. and Zentmire, Zelma. "The Prevalence of Vitamin A Deficiency among Iowa Children," *Journal of the American Medical Association.* Vol. 106. March 21 1936; "A Clinical Method for Determining Moderate Degrees of Vitamin A Deficiency." *Journal of the American Medical Association.* Vol. 102. March 24 1934; Jeans, Philip C. "Vitamin D Milk, the Relative Value of

evidence of vitamin A deficiency; Lydia Roberts at the University of Chicago is doing research on vitamin C and Palmer is concerned with the effect of undernourishment on night-blindness. While the techniques developed by these researches may be too difficult to permit of general applicability, a student of detailed nutrition problems should be familiar with them.

The relationship between undernourishment and derivative effects such as disease, school attendance and retardation, employability and similar problems is not clear. Evidence of abnormality in any of these fields can be traced to nutrition faults only with the greatest difficulty because of lack of knowledge and because numerous other factors such as age, epidemics, loss of skill among unemployed, and larger classes in school complicate the picture. While some studies along these lines may be fruitful, it is likely that considerable methodological experimentation is called for.

It has already been shown that nutrition studies demand the control of many factors; among them are normal differences in nutrition habits, weight, and standard of living among members of different economic groups; differences in liberality of relief grants and methods of distributing relief; and changes in the purchasing power of the dollar. The latter is particularly important in connection with cost of living studies. Wage earning families with a steady and unvarying income may be better nourished during depression than normal times. Similarly, time studies of food expenditures may be invalidated unless they reveal quantity purchases.

This extended recital of difficulties besetting the investigator of problems of nutrition in the depression should not deter the social scientist for he has probably never undertaken a problem that did not have controls as numerous or as complicated as

Different Varieties of Vitamin D Milk for Infants: A Critical Interpretative Review." *Journal of the American Medical Association*. Vol. 106. June 13, 1936 and June 20, 1936

these. Despite them a number of significant researches can be suggested.

(1) *Trend Study of Weights of School Children.*—A trend study, covering the business cycle, of the weight of children from records kept by the schools. Age (or school grade), sex, and, if possible, height should be controlled and there should have been no marked change in the racial or nativity composition of the school population or comparisons will be vitiated by changes in body build. This difficulty will be avoided if identical children can be followed over a period of years with observations of either actual weights or gains in weight. In either study control groups would be obtained by separating the data according to economic or employment status of the families. In the absence of a better index, the occupation of the father as recorded in the school records would serve as a rough measure of economic status. These studies can be undertaken on a small scale in one community or many similar studies over a wide area might be conducted according to a standard procedure. In the latter case, it is important to include some communities where relief was adequate and others where it was inadequate.

(2) *Trend Study of Physical Examinations of School Children.*—A similar study using simple physical examinations instead of weight. Physical examinations of school children are quite common and records over a period of years may be found. The ratings of nutrition status would presumably have been based on estimates of physical condition, amount of subcutaneous fat, muscle tone, skin color, etc. It is important to recognize in a study of this type that these measures may be highly subjective, and that a change of medical examiners may invalidate the data. The economic controls and differentiations mentioned in the first proposal are contemplated.

(3) *Physical Examinations of Employed and Unemployed Workers.*—A study based on physical examinations of workers unemployed for varying periods compared with examinations of

workers continuously employed. There are probably few records. Data on the unemployed, however, can be found in the offices of the United States Employment Service and on employed persons at the place of employment. Age and occupation as well as nativity should be controlled. An example of a study in this field is that by Diehl[93] who compared physical examination results for the employed and unemployed.

(4) Study of Fatigue among Workers Recently Unemployed.—Study 3 might be combined with another designed to show the effects of undernourishment on employability. Production records and fatigue might be measured for workers returning to work after periods of unemployment and compared with the records of those who were not unemployed. Several difficulties would have to be surmounted. Production and fatigability may be functions of habit and of the retention of skill as well as of physical condition. Also the workers with poor records were almost certainly among the ones laid off.

(5) Study of Gains in Weight among School Children Living Under Different Relief Policies.—A study of weight records of school children, using gains in weight of children from families receiving commodity relief compared with those from families receiving food orders and cash grants would reveal information about the effect of different kinds of relief policies. Similarly, if weights of infants from relief families are available in any clinics, it might be possible to arrive at a gross estimate of the effect of providing fruit juice and cod liver oil to infants.

(6) Physical Examination of School Children and Relief Policies.—A study conducted over a period of time and using school physical examination and weight records of low-income families differentiated on the basis of receipt of assistance and non-receipt might be expected to show something of the effect

[93] Diehl, H. S. "Physical Condition and Unemployment." *Public Health Reports.* November 15, 1935. Reprint No. 1716

of relief on nutrition among families of this group.

(7) Histories of Cases of Maternal and Infant Mortality.— The economic histories (and nutrition histories, if possible) of cases of maternal and infant mortality can be collected and compared with similar data for cases having successful outcome. The comparison of fatal with non-fatal cases is of vital importance.

(8) Weight of Infants at Birth.—Weight and other measures of the nutrition of children at birth can probably be obtained over a period of years from hospital records. Nativity, economic status, relief status, and similar factors should be controlled.

Chapter III

Environment and Health

IT HAS been shown in Chapter II that there is an association between income and sickness. Perhaps the most serious limitation of the studies suggested is that ill health is not a result of low income itself but of more fundamental factors such as inadequate supplies of food and clothing, poor housing, lack of medical care, and the conditions under which people work for a living.

It has been found, for example, that pellagra among low income cotton-mill workers is caused, mainly, by the absence from the diet of green and leafy vegetables, lean meats, fresh fruit, milk, butter, and eggs during certain seasons of the year.[1] Lack of facilities for the sanitary disposal of human excreta coupled with the practice of going barefoot gives rise to the spread of hookworm.[2] In earlier days typhoid was frequently spread through milk, water, and food supplies. It is not unlikely that such infections occurred more frequently among the low income groups than among the well-to-do. While the incidence of occupational diseases and the frequency of industrial accidents are closely related to the use of precautionary measures and safety devices, it is believed that less well defined factors such as hours of work, noise, and speed of movement also influence the vitality of the worker.

---

[1] Goldberger, Joseph, Wheeler, G. A., Sydenstricker, Edgar, King, Wilford I., *et al.* "A Study of Endemic Pellagra in Some Cotton-Mill Villages of South Carolina." *Hygienic Laboratory Bulletin.* No. 153. P. 63. Washington: Government Printing Office. January 1929

[2] Sydenstricker, Edgar. *Health and Environment.* A Committee on Recent Social Trends Monograph. New York: McGraw-Hill Book Company. 1933. P. 56

Unfortunately, research of the kind suggested by these examples is highly complex and has not been carried far, with the result that little of a precise nature is known concerning the amount of ill health and the specific environmental factors related to it. Thus, while the depression brought many changes in both living and working conditions, it is exceedingly difficult to set up studies that will show either (1) the relationship between health and any *specific* environmental factor or (2) the effect on health of a change in the fundamental environment. However, in view of the great importance of living conditions on human vitality and of the profound changes that must have taken place in many of those conditions, it is considered worthwhile to include a chapter on environmental changes, even though they cannot be used as indexes of probable specific effects upon the health of the people affected. The present chapter, therefore, is devoted to a brief discussion of: (1) minimum income for subsistence (2) consumption of housing, food, and clothing (3) occupational environment and (4) certain aspects of what may be called the social environment. In considering these topics their general as well as their depression phases are discussed. The receipt of care for the prevention and treatment of sickness is, of course, one of the most important of all environmental circumstances. Because medical care is designed solely for the promotion of health, discussion of it has been reserved for a final chapter.

MINIMUM INCOME FOR SUBSISTENCE

The demonstrated relationships between health and income are sufficiently close and sufficiently general to justify, as has been pointed out in the foregoing discussion, an approach to health problems through a study of the change in income itself. Specifically, one might set up as a standard some minimum budget and ask the question, "At given periods of time, during prosperity, depression, and recovery, what proportion of the families of the nation was not able to maintain such a budget?"

A. Determination of Minimum Subsistence Standards

The determination of a standard minimum budget is not, of course, a scientific problem alone, but is also a problem in human welfare. What is meant by an adequate minimum standard of living depends on the concept of adequacy. Opinions will differ. The concept of adequacy applied to the budget of a Negro share cropper by one familiar with Southern conditions would doubtless differ from the concept of adequacy as applied to the workingman's budget in a Northern industrial city. Perhaps the easiest component of a minimum budget on which agreement might be obtained most readily is food. Yet our knowledge of minimum food requirements to maintain health is still far from satisfactory, and extensive research in this field is needed. Notable among the efforts made to set up a minimum subsistence standard is the estimate by the Works Progress Administration of quantity requirements for basic maintenance and emergency standards of living for persons of different ages and sex.[3] The purpose and scope of these estimates are outlined by the author, Margaret Stecker, as follows:

> The material in this bulletin is the result of an effort to set up a technique for determining the cost of maintaining an adequate standard of living at the lowest economic level, and to establish quantity estimates of goods and services necessary to maintain that standard, on the basis of which costs at an identical standard in different localities may be compared. Because of the economic situation prevailing during the period within which this budget was constructed, an attempt was also made to ascertain how cuts below this basic maintenance standard may be made under emergency conditions, with least harm to individuals and the social group. While the approach to this study has necessarily been from the standpoint of relief, the resulting budgets are applicable generally, with

[3] Stecker, Margaret Loomis. "Quantity Budgets for Basic Maintenance and Emergency Standards of Living." *Research Bulletin*. No. 21. Series 1. Washington: Works Progress Administration. 1936

little or no modification, to low-cost living in urban areas, and should be of service in any field where information of this nature is required.

The budget content and other bases for computing costs are set up with reference to the needs of individuals living in a family group. These needs are related to the following persons: male industrial, service or other manual worker of small means; his wife who does all the work in the home, including cooking, cleaning, laundry, etc.; children of both sexes, between the ages of 2 and 15, inclusive. Individual requirements, in turn, can be combined for family groups of any size and composition within the limits set. No provision is made in the budget for estimating costs for individuals who live apart from a family group. Most items listed and their quantity weights are of general application in urban areas throughout the country; such special accommodations as are necessary result from differences in climatic or other local conditions, not from differences in standard of living or individual needs.[4]

The rather comprehensive minimum budget developed by the Works Progress Administration is strictly applicable only to urban communities. Perhaps the most adequate factual information for determining how such a minimum budget should be modified to take into account rural-urban differences, as well as climatic differences, should be forthcoming from the Study of Consumer Purchases made in 1936 by the Bureau of Labor Statistics in cooperation with the National Resources Committee, the Works Progress Administration, and the Department of Agriculture.[5] Information for this study was gathered by the house-to-house canvass method in representative districts throughout the United States and includes reported incomes and detailed expenditures of families in urban, rural, and semirural communities. After one has determined from the Study of Consumer Purchases, and other sources, the weights which need to be assigned in different regions to items selected as essential in a minimum budget, one can then determine the amount of money which would have been necessary, at various stages

[4] *Ibid.* P. 1

[5] "Plans for a Study of the Consumption of Goods and Services by American Families." *Journal of the American Statistical Association.* 31:135-140. March 1936

of prosperity, recovery, and depression, to purchase those items. The indexes of the Bureau of Labor Statistics and the basic data from which these indexes were derived should enable the research worker to secure a close approximation of the fluctuating cost of the selected budget items.[6]

B. Incomes Below Subsistence Standard

If one has available a standard, in the form of the income necessary for minimum subsistence in various years, one can then seek to apply this standard to any population for which an estimate of income distribution can be made.

Practically, it is doubtful whether enough information ever can be obtained pertaining to the period 1930-35 to provide an adequate estimate of the frequency distribution of incomes for each year.[7] A complete frequency distribution of incomes is, however, not necessary. What is necessary is information which will give at least a rough indication of the proportion of incomes below the given arbitrary amount selected as that which

[6] A more detailed discussion of these problems possibly would be outside of the scope of the present monograph. The reader is referred to Vaile, Roland S. *Research Memorandum on Social Aspects of Consumption in the Depression.* (monograph in this series)

[7] In the absence of direct reporting, economists, in the past, have had to resort to indirect methods of computing the amount and distribution of incomes in this country. For the past fifteen years the National Bureau of Economic Research has been collecting masses of data in an effort to arrive at the annual national income and its distribution. The Brookings Institution, making use of the publications of the Natural Bureau of Economic Research, supplemental information from other sources, and original investigations of its own, prepared a distribution of income for 1929. See Leven, M., Moulton, Harold G., and Warburton, Clark. *America's Capacity to Consume.* Washington: The Brookings Institution. 1934. Estimates of the gross national income from 1929 to 1935 have been made by the Bureau of Foreign and Domestic Commerce. See *National Income, 1929-32.* Senate Document No. 124, 73rd Congress, 2nd Session. Washington: Government Printing Office. 1934. Also, *National Income of the United States, 1929-1935.* Washington: Bureau of Foreign and Domestic Commerce, U. S. Department of Commerce. 1936

is necessary for minimum subsistence. However, even such information cannot be had accurately. The only question is whether it can be had with sufficiently small error to provide at least a rough basis for estimating the changing proportion of families living at a substandard level.

Attention of economists specializing in this aspect of the field should be called to the interest which the student of health in its social aspects would have in income estimates which contain not too large a margin of error. In the present monograph no more can be done than to sketch very briefly some of the possible lines of attack. The problem may be considered in two parts—first, that of securing information for the period of the depression of the 1930's, and, second, that of laying the basis for reporting in future years. Let us consider first the problem of securing information for the depression of the 1930's on the proportion of the population with an income below some designated standard.

The families in the lowest income groups may be divided into three classes, namely, those on relief, those not on relief but receiving emergency government aid in some form, and those neither on relief nor receiving other emergency government aid. These groups might be sub-divided again into farm and non-farm. The income of non-farm families on relief or in receipt of government subsidy in the form of emergency employment of the breadwinner (for example, in WPA), probably could be estimated in such a way as to determine the proportion of families in this group with an income below an indicated standard. This group will be considered in more detail later in this section. The proportion of farm families falling below the indicated standard might be estimated by use of the data on gross farm income in the *Census of Agriculture*, 1930 and 1935[8] supplemented by other information available in the

[8] Volume III, County Table III of the *Census of Agriculture*, 1930 gives farm income in terms of value of products sold, traded, or used by the opera-

records of the Department of Agriculture and the Rural Resettlement and the Works Progress Administrations. Possibly the greatest difficulty would arise in attempting an estimate of the proportion of non-farm families not in receipt of relief or other emergency government subsidy who fell below the indicated standard. Whether the number in this group is large or negligible depends, of course, on how low the standard is set. If the standard should prove to be $1,000 annual income for a family of average size, it may be that the number of such families not receiving any form of relief or government subsidy is quite large. Preliminary tabulations of the Health Inventory for one Northern industrial city indicate that 45 per cent of all the families in this city had an income of less than $1,000 in 1935, and of these families only about one out of three was receiving relief or benefiting through Works Progress Administration jobs.[9] Whether or not it would be possible to estimate the number of such nonrelief families in non-farm areas in the country as a whole for earlier years, the writers are not prepared to say. But it is possible that an error of twenty or thirty per cent in this estimate might not produce a very large error in the aggregate estimate, provided the figures on relief families in non-farm areas and of all low-income farm families could be somewhat more accurately estimated.

One hope for future information on income distribution is from direct house-to-house inquiries. For the year 1935, the data of the Health Inventory should provide the most extensive direct information on the distribution of incomes ever obtained

---

tors' families in each county in the country. These data can be used to estimate the proportion of farm operators with incomes below a specified amount. The *Census of Agriculture,* 1935 lacks published detailed information of this sort, but the Department of Agriculture has made annual estimates since 1924, for divisions and states for owner-operators only, but not for counties. See *Agricultural Statistics. 1936.* U. S. Department of Agriculture. Table 447

[9] From preliminary tabulations of the National Health Inventory

in the United States.[10] The experience of this survey, together with that of the Study of Consumer Purchases of 1936, and the Michigan Census of Population and Unemployment[11] of 1935 should provide a valuable pool of information to guide those who would seek to determine the distribution of incomes in the United States in future years. There have been frequent suggestions that the Bureau of the Census include income as an item on the census schedule. One reason for omitting this extremely important entry has been doubts as to the reliability of the reports, although in some other countries, including Canada, the information obtained has seemed plausibly accurate. It might be especially desirable to make cross-tabulations of occupation, rental, and income from the schedules of the three large surveys mentioned above. If there is a consistent correlation, squaring with expectations as to the relationships between these items, one would have increased confidence in the validity of the income reports. No cross-tabulations of rental and income have yet been made by the Health Inventory, but the preliminary cross-tabulations of occupation and income conform remarkably well with a priori expectations.

It hardly need be mentioned here that, in attempting to determine the proportion of families whose income falls below a designated minimum, allowance must be made for size and composition of the family. The degree of refinement that is advisable depends, of course, on the degree of accuracy of the

[10] In 1934 a work project sponsored by the California State Medical Society obtained estimates from several thousand families of annual income in both 1929 and 1933. The data for 1929 follow the estimates of the Brookings Institution with remarkable fidelity, and should not be overlooked in the present connection. The study was directed and the report (as yet unpublished) *Economic Aspects of the Practice of Medicine in California,* was prepared by Professor Paul A. Dodd of the University of California.

[11] *Michigan Census of Population and Unemployment, No. 6.* "Total Income During 1934 of Gainful Workers." Lansing: State Emergency Welfare Relief Commission. March 1937

original figures. For very broad estimates, size of family alone might be sufficient. For refined estimates, such as might be made for a local region on the basis of data collected in the Health Inventory, it would be desirable to introduce weights and reduce all figures to an adult-male equivalent or ammain basis. Eventually more adequate data on income may be available through the records of the Social Security Board or through the Internal Revenue Bureau if a broader income tax base is adopted.

c. Adequacy of Incomes of Relief Families

Subsequent research may find too many elements of uncertainty to justify generalizations—even with a rather large margin of error—about the proportion of all American families in the depression with income below a minimum subsistence standard. The effort should not be given up until every reasonable possibility of success is thoroughly exploited. Meanwhile, however, there would be much value in generalizations about the extent to which relief families in themselves attained or fell below a minimum subsistence standard. Here is a major problem for which there is an abundance of data awaiting analysis.[12]

At the peak relief load some five million heads of families representing about twenty million individuals of all ages were dependent upon public relief.[13] It has been intimated above that, during the time they were receiving relief, they probably received goods and services insufficient to maintain them in good physical condition. On the other hand, there is some evidence that the condition of the chronic poor among them actually improved under depression relief.[14] It is certain, however, that there is a

[12] Further discussion of this subject for research appears in the monograph in this series by White, R. Clyde and Mary K. *Research Memorandum on Social Aspects of Relief Policies in the Depression.*

[13] *Monthly Report of the Federal Emergency Relief Administration January 1 through January 31, 1935.* Washington: Government Printing Office. 1935

[14] Wiehl, D. G. "Diets of Low-Income Families Surveyed in 1933." Health and

wide variation in income accruing to individuals among families of different sizes and compositions existing upon work relief wages when these incomes are distributed according to the number of "adult male units" deriving their sustenance from them.

A study of living standards of relief families should cover predepression as well as depression years. There should be included a sufficient sample of outdoor relief provided in predepression years to establish the trend of the volume of dependency occurring independently of cyclical changes in industrial activity, and of the quality of relief ordinarily afforded. Examination into this phase of the problem may prove it to be impractical due to the diversity of the agencies providing outdoor relief prior to the depression, but its importance in estimating the effect of the depression as an instrument in creating destitution still persists.

Relief during the emergency years already experienced, 1930-37, passed through several phases, each of which will serve to complicate any investigation. Throughout the entire period several types of relief were provided, such as direct relief in kind, direct money relief, work relief, and a combination of all three. The administration of relief also passed from private to public agencies as the depression deepened, and the public agencies administering relief changed in some instances from local to state administrations. Case histories of relief families, however, were frequently passed from agency to agency and it is believed that unbroken sequences occur in many localities.

The research data divide themselves rather naturally into three periods and as many phases. The first period extends from 1930 through May 1933 and is characterized by direct relief locally financed and administered. The second period extends from May 1933 through October 1935 and is characterized by a combination of direct relief and work relief as furnished by

Depression Studies No. 3. *Public Health Reports,* January 24 1936. Reprint 1727. P. 4

the Federal Emergency Relief Administration throughout the entire period and by the Civil Works Administration from November 1933 to April 1934. Over 70 per cent of this relief was financed by federal grants in aid administered by state relief committees.[15] The third period extends from the last half of 1935 to date and is characterized by work relief furnished and supervised largely by the federal government for employable relief clients and by direct relief through local governments for unemployables.

Since it is the purpose of the investigation here suggested to determine the annual income of each individual subsisting from relief, and since records show that a complete numerical turnover in the relief rolls occurred approximately every two years, it would seem desirable to conduct any investigation on an annual basis. That is, relief rolls in selected urban and rural communities could be sampled by each year of the depression.

Direct Relief.—Case records of many relief agencies may be relied upon to furnish accurate data on the value of relief received by each family from public sources. This does not represent, however, the entire income of all families on relief. In addition to direct relief payments in money and kind many relief clients supplemented their incomes from one or more of the following sources: (1) loans of money (2) credits in the form of unpaid house rents and unpaid commodity purchases (3) low paid or part time jobs (4) reserves in the form of insurance, liquidation of personal and real property, and soldiers' bonus and (5) gifts of food, clothing, fuel, and so forth from nongovernmental sources.

Case workers attempted to appraise the value of these increments to family incomes and to adjust the amount of relief accordingly. It is not to be considered, however, that the case

[15] *Monthly Report of the Federal Emergency Relief Administration, July 1 through July 31 1935* and subsequent numbers. Washington: Government Printing Office. 1935

records will begin to reflect the total value of these types of income. Studies of the family histories of relief applicants and Civil Works Administration workers such as the Clague-Powell report[16] and surveys made by the Federal Emergency Relief Administration will be useful as a guide in determining the value of such sources of revenue.

Work Relief.—Local, state and federal work relief programs were designed to substitute useful employment for the dole. Wage scales were generally fixed at some point sufficiently below prevailing rates to encourage relief workers to seek private employment but at the same time sufficiently high to provide the recipients with all of the bare necessities of life.

So called "security" wages varied horizontally with the type and size of the community and vertically according to occupational classification. The lowest rates were paid to unskilled workers in rural communities and the highest wages to professional and technical workers in larger cities.

Since work relief was allowed to only one worker per family, income accruing to the family was fixed without regard to variations in size and composition. While hourly rates paid to the lower occupational classifications were low the continuity of employment was often better than private industry afforded and insured a steady fixed income against which family expenditures could be budgeted.

Income from work relief for single persons and small families, undoubtedly, was sufficient to provide these recipients with relatively healthful living; it is doubtful if very large families which had to rely upon work relief for sustenance were able to secure an adequate diet, adequate clothing, sanitary housing, and medical care. An inquiry, therefore, into the incomes of the various kinds and types of families receiving work relief will be of value.

---

[16] Clague, Ewan and Powell, Webster. *Ten Thousand Out of Work.* Philadelphia: University of Pennsylvania Press. 1933

Records of wage payments to relief workers as made by local and state agencies will be carried in most offices on the case cards of each worker. When relief clients were certified to the Works Progress Administration, however, their cases were closed as far as the local relief agency was concerned and records of wages received will have to be taken from Works Progress Administration payrolls. These rolls are conveniently located in the various Works Progress Administration Area Statistical Offices.

A study of the case records and Works Progress Administration payrolls of a sample of the relief population should provide the following data: (1) distribution of families according to size and composition (2) monthly income (3) usual occupation of head or employed member (4) location, type and size of family dwelling and rent paid (5) medical care furnished and (6) period of dependency.

This information, if records are accurately kept, will provide all of the data necessary to compute the range of relief income of these families when reduced to "adult male units" and to determine the adequacy of relief to provide a minimum subsistence standard of living.

The information from these sources does not provide, of course, the elements necessary to interpret results directly in terms of associated mortality, morbidity, or physical impairment. It does provide, however, a measure of the relative amount of exposure to known unhealthy conditions of living among an important segment of the population which in normal activity might or might not be subject to ailments in excess of those displayed by the general population.

D. Summary

This section has sought to indicate that the relationship between income and health justifies research which focuses not on health in itself, but on the extent to which the population lacked the minimum income deemed necessary for adequate subsistence. This implies two general types of inquiry: (1) into the problem of determining minimum standards, and (2) into the problem of finding what proportion of the general population of a region or of the country as a whole fell below that standard. Even if the second line of inquiry, because of the numerous difficulties inherent in the source material, proves unsatisfactory, it would still be feasible and valuable to make this inquiry with respect to the relief population, whose characteristics are rather well known from numerous studies.

CONSUMPTION

Estimates of family income would provide fairly good evidence of the reality and extent of inability to purchase the goods and services necessary to maintain health. This would be particularly true if the estimates could be seen in relation to budgetary standards. Such evidence, however, does not reveal the places in which shortages may have occurred. It is believed that sufficient data are available to give some conception of what may have occurred in the consumption of housing, food, and clothing. It is quite possible that consumption may have been restricted not only through actual inability to make purchases, but also through reluctance on the part of many with sufficient resources to spend money because of a feeling of insecurity. That this may have given rise to an actual shortage of commodities even as late as the recovery period appears to be true in the case of housing. The discussion which follows is limited because the topic of consumption is the subject of another monograph in the series.[17]

A. Housing

Housing and Health.—The allowance for housing in the quantity budgets is based upon the rental value of existing dwellings having physical attributes considered essential to

[17] See Vaile, Roland S. *Research Memorandum on Social Aspects of Consumption in the Depression.* (monograph in this series)

health.[18] It is generally known that there are excessive morbidity and mortality rates among infants and among older persons who reside in substandard dwellings. To what extent these excesses are due to housing and to what extent they are explained by other factors has not been determined. Britten, however, has set down several hypotheses concerning the supposed relationship between housing and health. The particulars are as follows:

(1) Lack of pure water through (a) the use of wells and cisterns in slum areas and (b) the non-piping of city supplies into slum dwellings, offers risks of epidemic diseases.

(2) Insanitary sewage disposal by means of (a) privies which are improperly constructed and maintained, (b) inadequate toilet facilities requiring multi-family use, and (c) privies and toilets poorly located in yards, porches, and halls, may be responsible for the prevalence of dysentery, infant summer complaint, and hookworm.

(3) Congestion and overcrowding of sleeping and living quarters, halls, yards, and streets facilitate the spread of communicable diseases, such as the common cold, influenza, chicken pox, cerebro-spinal fever, diphtheria, and cholera.

(4) Absence of ventilation or poor ventilation, both of which conditions obtain in the slum, play some part in the occurrence of such diseases as pulmonary tuberculosis and pneumonia.

(5) Inadequate opportunities for securing sunlight in slum homes and districts probably have some relation to the prevalence of rickets among children living there.

(6) Slum houses are not usually screened or at the most poorly screened against flies which are purveyors of typhoid fever and diarrheal diseases, and mosquitoes which are responsible for malaria. In addition, slum dwellings are very seldom rat-proof whereas plague is endemic among rodents in certain parts of the country.

(7) Extreme dampness—flooded cellars, leaking roofs, moulding walls, etc.—may be associated with rheumatism and the like.

(8) Dilapidation, i.e., broken stair steps, missing or uneven floor boards and so forth mean an increased risk for home accidents. What is more, the advanced age of most of the dwellings in the slum means that they are seldom fireproof and that the hazard from death and injury by fire is therefore very great.

[18] Stecker, Margaret Loomis. "Quantity Budgets for Basic Maintenance and Emergency Standards of Living." *Research Bulletin.* No. 21. Series 1. P. 28. Washington: Works Progress Administration. 1936

Up to this point the effect of slum areas on health has been measured in terms of mortality or sickness rates; but it is to be recognized that health embraces more than the mere absence of outright disease; it is a state of being in which all physical and mental processes function at their highest efficiency. Influences which affect physical or mental efficiency, or the peace and comfort of the family, are therefore to be regarded as having an adverse effect on health. It is clear that most of these influences are intangible, and are so bound up with poverty as such, with the worry of unemployment, with limited facilities for medical care, with lack of cleanliness, that we should not expect their elimination by the demolition of a given slum area and the rehousing of its inhabitants. . . .[19]

The Depression Strikes Housing.—If it can be assumed that a family ordinarily lives in as good a dwelling as it can afford, then the general curtailment of income must have forced many families to accept poorer quarters than they had been occupying prior to the depression. This must have been the case even when the necessary allowance is made for reduced rentals, rent paid by relief agencies, and non-payment of rent. Efforts at adjustment probably assumed a number of forms: (1) failure to build new dwelling units in keeping with the increase in population and the corresponding failure to demolish old units (2) doubling up of families (3) acceptance of poorer quarters and (4) failure to make necessary repairs to structure or equipment. Concrete illustration of the manner in which adjustments are made is found in the following paragraphs quoted from a United States Department of Labor publication, *Earnings and Standard of Living of 1000 Railway Employees During the Depression:*[20]

The lowered standards of living (of 1,000 railway employees during the depression) are also reflected in *housing changes.* Two hundred and twenty-five families had moved at least once within the last 4 years, mainly to reduce their rent. Some families reported 3 and 4 changes in

[19] Britten, Rollo H. "Relation between Housing and Health." *Public Health Reports.* November 2, 1934. P. 1308

[20] Goodrich, Carter. *Earnings and Standard of Living of 1000 Railway Employees During the Depression.* Washington: Bureau of Labor Statistics, U. S. Department of Labor. 1934. P. 27

this period, each time to less expensive quarters. At the time interviewed 39 per cent of the families who rented their homes, and 9 per cent who owned or were buying them in 1929 were living in other quarters than those occupied in 1929.

In some instances these changes entailed only a less convenient or less pleasant neighborhood, but in half the cases the changes were into smaller, inferior, poorly equipped houses or apartments. Families accustomed to furnaces and hot-water facilities were living in "cold flats," often without electricity or running water. To give up inside toilets and baths, as many families had to, seemed like a reversion to barbarism. Cheap rents are to be found in noisy, dirty, crowded streets, where children must take their recreation and may come in contact with undesirable playmates. Families who moved in together or into smaller apartments complained bitterly of the lack of privacy. In one case 3 families (7 people) had moved into 4 small rooms opening onto a congested court shared by 4 other householders.

Fuel was another item sacrificed in cutting housing costs. Practically all the families economized to some extent in this respect, but 53 stated that this was one of their chief economies. . . . individual burners were used in place of the furnace. All the rooms that could possibly be spared were being shut off to avoid heating them. Often wood was used instead of oil for cooking. According to reports, discarded railroad ties, hardly a satisfactory kind of fuel, were at a premium. . . . in some families the need for bedding, linens, and so forth was acute.

As may be supposed, most of the families owning property reported that even necessary repairs were being postponed. Many men would have done this work themselves in their spare time had they had the funds for the necessary supplies.

Quantitative estimates of what happened to housing in the United States are given in Table I.[21] From this table it may be seen that, while the number of potential occupants (private families) continued to increase throughout the depression, the number of vacancies also increased throughout its downward swing despite a practical cessation in building activity. It appears from these data that many families were forced to double up and that others must have occupied dwellings which in normal

[21] See also the article by Foster, R. R. and Wickens, David L. "Estimating the Volume of Residential Building Construction," *Journal of the American Statistical Association*. 32:97-104. No. 197, March 1937

TABLE I

NUMBER OF FAMILIES AND NUMBER OF DWELLING UNITS BY TYPE OF OCCUPANCY: NON-FARM UNITED STATES, 1921–1936[a]

Year	No. of Private Families (in Thousands)		Number of Dwelling Units (in Thousands)						
	Total Dec. 31	Net Increase During Year	Standing Dec. 31	Built During Year	Demolished During Year	Net Increase Standing During Year	Occupied Dec. 31	Abnormally Occupied by Two or More Families[b] Dec. 31	Vacant Dec. 31
1921	18,685	580	18,761	446	19	427	18,485	200	276
1922	19,318	633	19,435	705	31	674	19,143	175	292
1923	20,076	758	20,265	846	16	830	19,951	125	314
1924	20,622	546	21,099	872	38	834	20,572	50	527
1925	21,203	581	21,937	951	113	838	21,203	0	734
1926	21,804	601	22,683	845	99	746	21,804	0	879
1927	22,306	502	23,409	781	55	726	22,306	0	1,103
1928	22,729	423	24,093	736	52	684	22,729	0	1,364
1929	23,200	471	24,539	472	26	446	23,200	0	1,339
1930	23,491	291	24,763	260	36	224	23,291	200	1,472
1931	23,646	155	24,945	200	18	182	23,196	450	1,749
1932	23,674	28	24,982	80	43	37	22,974	700	2,008
1933	23,952	278	25,011	70	41	29	23,102	850	1,909
1934	24,388	436	25,027	60	44	16	23,588	800	1,439
1935	24,849	461	25,125	150	52	98	24,149	700	976
1936	25,316	467	25,350	200	75	225	24,716	600	634

[a] Estimated from unpublished data loaned by Mr. George Terborgh, Washington, D.C.
[b] The number of families living with other families is, of course, at least double these numbers.

times would have been destroyed. By 1934 families that had moved in together at the depth of the depression began to separate. This fact, together with the relative stability in the number of standing units from 1932 to 1934, may have left the nation in 1936 with a more acute housing shortage than existed in the years immediately following the war.

Other evidence of continued use of old housing facilities may be derived from the *Biennial Census of Manufactures*[22] which

[22] *Biennial Census of Manufactures.* Washington: U. S. Bureau of the Census. Vols. for 1929, 1931, 1933.

provides information concerning the fabrication of plumbing supplies including bathtubs, lavatories, sinks, closet bowls, flush tanks, and other equipment.

It is readily seen that adjustments along the above lines must have had implications for the health of many of the families involved. While it is probably impractical at this date to trace the movements of even a small number of families from one dwelling to another as they attempted to cope with reduced income and to describe the units successively occupied or to measure deterioration of specific units, it is believed that studies along the lines suggested above can be used to arrive at some approximation of changes in housing conditions.

TABLE II

NUMBER OF FAMILIES AND DWELLING UNITS IN DETROIT: 1928–1936

Year	Number of Private Families[a]		Number of Dwelling Units[b]			
	Total	Net Increase	Standing	Built	Demolished	Net Increase in No. Standing
1928	375,920	+11,884	383,000	15,929	355	+15,774
1929	377,260	+1,340	394,879	12,149	270	+11,879
1930	365,640	−11,620	398,557	4,084	406	+3,678
1931	350,165	−15,475	400,289	2,135	403	+1,732
1932	345,680	−4,485	400,128	310	471	−161
1933	354,700	+9,020	399,596	265	797	−532
1934	371,150	+16,450	399,103	411	904	−492
1935	391,980	+20,830	400,000	1,647	750	+897
1936	405,000	+13,020	403,941	4,926	985	+3,941

[a] Number of private families derived from annual population estimates of the R. L. Polk Co.

[b] Number standing, 1935, number built, 1928–36, and number demolished, 1928–36 from reports of Department of Buildings and Safety Engineering, City of Detroit; other dwelling unit figures derived from these.

Estimates of home construction and occupancy similar to those presented above for the United States can be made for nearly all cities of any size. Most places keep official records of construction and demolition. Other local agencies such as the real estate board, the post office, the electric and gas utilities, and the telephone company often can provide estimates of vacancies. Intercensal population estimates are more difficult to obtain but sometimes are available from local censuses or family directories. With data such as these, Table II was constructed for the city of Detroit. It appears from these data that Detroit has had an excess of housing over the number of private families, but that restriction of building activity during the depression finally brought about an actual shortage. Recent data indicate that while there was an estimated population decrease of 13 per cent between 1929 and 1932 in this city, the number of vacant dwelling units increased 30 per cent.[23]

It is common knowledge that during the depression many repairs were not made because even the relatively minor costs could not be afforded. Unfortunately, there is no satisfactory method of obtaining estimates of trends in the volume of repairs made. However, some notion of the magnitude of change may, perhaps, be derived from indexes of major alterations. These can be derived from the records of permits for remodelling and for plumbing work. Many cities require permits for these purposes and maintain records of the number issued.

B. Clothing and Household Necessities

Consumption of various articles of clothing such as shoes, dresses, overalls, suits, gloves, and hosiery, and of household goods such as stoves, cooking utensils, screens, window glass, and bedding should be considered in the present connection even though the relation between adequacy of clothing and health status is less definitely known than that between housing and health. The *Biennial Census of Manufactures*[24] and the *Statistical Abstract of the United States*[25] provide data that may be employed to arrive at gross estimates of consumption of the items

[23] *The Detroit News.* May 23, 1937. P. 14

[24] *Biennial Census of Manufactures. Loc. cit.*

[25] *Statistical Abstract of the United States.* Washington: U. S. Department of Commerce. Bureau of Foreign and Domestic Commerce.

mentioned above. An example of the possible use of these data is as follows: the quantity budget set up by the Works Progress Administration estimates that, on the average, three pairs of shoes are required annually for each individual in the population.[26] Table III shows the estimated number of pairs of shoes available in the United States from 1919 to 1933 together with the population and the number of pairs available per capita. The last two columns in this table indicate that except in the

TABLE III

TOTAL AND PER CAPITA SUPPLY OF BOOTS AND SHOES, 1919–1933

Year	Total Available Supply (in Thousands of Pairs)[a]	Estimated Population (in Thousands)	Pairs per Capita	Index 1929 = 100
1919	309,604	105,003[b]	2.95	96
1921	277,897	108,208[b]	2.57	83
1923	344,586	111,537[b]	3.09	100
1925	317,763	114,867[b]	2.77	90
1927	363,209	118,197[b]	3.07	100
1929	373,926	121,526[b]	3.08	100
1931	317,467	124,974[c]	2.54	82
1933	350,616	125,770[c]	2.79	91

[a] Calculated from production data taken from the *Census of Manufactures* and import and export statistics from *Statistical Abstract of the United States*. Issues for 1922 through 1935, Bureau of Foreign and Domestic Commerce, U. S. Department of Commerce, Washington.

[b] *The World Almanac*, 1935. Reprint of Bureau of the Census Mid-Year Population Estimates.

[c] Truesdell, Leon E. "Population Statistics," *American Year Book, 1936*. New York, Thos. Nelson and Sons, 1937. Part Five, Div. XIV, Pp. 540–45

most prosperous years the total available supply of shoes in the United States did not equal the estimated requirements. Further refinement of the estimates may be introduced by classifying shoes as for men, women, and children. A major defect of these data is that they provide no indication of the inequality of distribution that must have existed. Similarly, it is quite possible that the quality of shoes and of other commodities declined during

[26] Stecker, Margaret Loomis. "Quantity Budgets for Basic Maintenance and Emergency Standards of Living." *Research Bulletin*. No. 21. Series 1. P. 17. Washington: Works Progress Administration. 1936

the depression. If this is the case then neither the production figures just cited nor estimates in terms of value of the product are adequate measures of the real changes that may have taken place. This subject is discussed in the *Research Memorandum on Social Aspects of Consumption in the Depression* by Roland S. Vaile.

c. Food

It is well known that the production of food declined during the depression. Whether or not the declines were restricted to those foodstuffs among which there apparently had been overproduction is important from the viewpoint of health. Food requirements according to the quantity budget at the minimum subsistence level are based not only upon calorie values but also upon the requirements of a balanced diet.[27] Stiebeling and Ward estimate the yearly quantities of various foods or groups of food needed per capita for the population of the United States as shown in Table IV.

Nutrition studies indicate that the foods most frequently deficient in the diets of low income groups are: milk, citrous fruits and tomatoes, leafy, green, and yellow vegetables. There are several possible explanations for this, among them being (1) ignorance of the importance of these items (2) their costliness as compared to other foods or (3) their unavailability in local markets due in part to lack of proper facilities for transportation and preservation.

Because of the lack of uniformity in the distribution among all population groups of the several items in the quantity budget, it is evident that estimates of the gross national supply of foodstuffs is a very rough criterion of the extent to which the depression affected the adequacy of diets. Nevertheless, because of the very direct relationship between food consumption and health, consideration of the influence of the depression on the diet of

[27] Stecker. *Op. cit.* P. 7

TABLE IV

ESTIMATED ANNUAL PER CAPITA FOOD REQUIREMENTS BY TYPE OF FOOD, AT VARIOUS DIETARY LEVELS[a]

Food	Unit	Restricted Diet for Emergency Use[b]	Adequate Diet at Minimum Cost[b]	Adequate Diet at Moderate Cost[b]	Liberal Diet[b]
Flour, cereals	pounds	240	224	160	100
Milk or its equivalent	quarts	155	260	305	305
Potatoes, sweet potatoes	pounds	165	165	165	155
Dried beans, peas, nuts	pounds	30	30	20	7
Tomatoes, citrous fruits	pounds	50	50	90	110
Leafy, green, and yellow vegetables	pounds	40	80	100	135
Dried fruits	pounds	10	20	25	20
Other vegetables, fruits	pounds	40	85	210	325
Fats (including butter, oils, bacon, salt pork)	pounds	45	49	52	52
Sugars	pounds	50	43	60	60
Lean meat, poultry, fish	pounds	30	60	100	165
Eggs	dozen	8	15	15	30

[a] Stiebeling, Hazel K. and Ward, Medora M., *Diets at Four Levels of Nutritive Content and Cost*, Circular No. 296, U. S. Department of Agriculture, Washington, 1933.

[b] The figures given in this table are computed from diets adapted to the needs of individuals of different age, sex, and activity groups (Table 6) and from the number of persons in each group as shown by the 1930 census of population. The quantities are those which should be delivered to the family kitchen. To convert them into production figures, suitable margins must be added to the different food groups to cover the unavoidable losses in harvesting, grading, storage, manufacture, or distribution.

even the whole community is of real significance. Consequently, it is suggested that records of the Department of Agriculture, reports in the *Statistical Abstract of the United States* and other sources should be employed to determine whether or not there were marked changes during the past several years in the production and consumption of the various classes of food products included in the recommended budgets. The Study of Consumer Purchases[28] currently being completed will provide some basis for estimating differential food consumption among different income, cultural, and residential groups.

[28] "Plans for a Study of the Consumption of Goods and Services by American Families." *Journal of the American Statistical Association.* 31:135-140. March 1936

D. SUMMARY

Studies are proposed in the use and construction of dwelling units and in the consumption of foodstuffs, clothing and household necessities in order to interpret the family income studies proposed in the preceding section, by revealing at what places in the quantity budget shortages were most acute. While data are frequently scarce and not always readily adaptable to these purposes, the subject matter warrants the payment of considerable attention to them.

OCCUPATIONAL ENVIRONMENT

The mortality of male workers is definitely higher than that of the general population or of females of comparable social-economic status.[29] This is especially true of unskilled and semi-skilled wage earners among whom mortality rates, when adjusted for age, are markedly higher than those among professional and business people.[30] This excess mortality can be attributed: (1) to exposure to industrial accidents, hard dust, and poisonous substances (2) to fatigue, noise, confinement, and the strain of hard physical work[31] and (3) to the condition and habits of the worker himself.

In the thirty years ending with the onset of the depression, industry in this country had passed through an era of efforts to improve working conditions through the installation of safety

[29] Dublin, Louis I. and Vane, Robert J., Jr. "Causes of Death by Occupation." U. S. Bureau of Labor Statistics *Bulletin* No. 507. Washington: Government Printing Office. February 1930; Sydenstricker, Edgar. *Health and Environment.* A Committee on Recent Social Trends Monograph. New York: McGraw-Hill Book Co. 1933. P. 25.

[30] Britten, Rollo H. "Mortality Rates by Occupational Class in the United States." *Public Health Reports.* 49:1102, No. 38, September 21, 1934

[31] Sydenstricker in *Health and Environment* states that "Genetic, industrial and social selection, economic status, social class and social attitude—all of these are possible factors (in predisposing wage earners to ill health) but even when these are taken into consideration and given considerable weight, it is difficult to shake off the conviction that occupational environment, in addition to known specific hazards, is an important factor in the health of the working population"—p. 143. See also Dublin and Vane. *Causes of Death by Occupation.*

devices, air conditioning, noise reduction, factory inspection, and education of the worker in practices of safety, all of which resulted in a marked downward trend in the frequency of accidents. During the depression there was evidence of even a further decrease in industrial accidents, and in Chapter II a study was proposed to show the full extent of this further decline. Two explanations of this depression phenomenon have been offered: (1) that the poorer workers who are more likely to have accidents were unemployed and (2) that hours of work were often shorter and hence there was less exposure to accidents.[32] The latter would apply equally to exposure to poisons and dusts and to the less specific working conditions. Other factors may have operated to prevent the accident rate from declining more than it actually did. It would not be surprising to find that the depression had given rise to a relaxation of safety standards and rigidity of factory inspection and to a tendency to use machinery when it was no longer safe to do so.

The present section is devoted to a brief outline of the movement toward bettering conditions of work and to suggestions for studies that would reveal whether or not there had been a partial abandonment of some of these gains and, if so, what some of the consequences may have been.

A. The Changing Environment

The last years of the nineteenth century were characterized by extreme absence of consideration for the worker. Wages were inadequate, hours of work were excessive, and the physical conditions of factories were often inimical to health. Mortality of workers was high and disabling accidents were frequent.[33] Employers were not required to assume responsibility for accidents

---

[32] Collins, S. D. "The Health of the Worker," *The Annals of the American Academy of Political and Social Science.* 184:37 March 1936

[33] *Statistics of Industrial Accidents in the United States to the End of 1927,* U. S. Bureau of Labor Statistics Bulletin No. 490. Washington: Government Printing Office. 1929

and the entire burden was thrown upon the injured or diseased workman and his family, except in the few instances of recovery from the employer by court procedure under common law rules. Coincident with the beginning of the present century government, employers, trade unions, and technological advances began to operate as forces to alleviate the more adverse conditions.

Government Control.—Early in the new century government undertook to fix upon the employer the financial responsibility for accidents. The first efforts were largely unsuccessful because responsibility and the amount of compensation were not well defined, but in 1911 states began to stipulate specific sums to be paid for specific injuries or as compensation for time lost. By 1930 all but three states had enacted compensation laws. These laws impose upon the employer all or most of the cost of medical care for the injured and a majority of them cover occupational diseases. One result of these laws was to make it profitable to the employer to prevent accidents through installation of safety devices and other plant changes.

Employers and the Profit Motive.—Efforts on the part of employers to reduce the cost of workmen's compensation imposed by law by reducing the frequency of accidents through safer conditions of work, soon demonstrated the fact that expenditures of this nature could be turned into profitable investments.[34] The productivity of labor in general was increased wherever conditions of work were improved. There were fewer plant stoppages due to accidents, less loss of time among individual employees and a lower labor turnover.[35] As a result of these favorable developments the larger users of labor voluntarily began to

[34] See Fisk, C. T. "What Safety Means to Buick," *National Safety News.* 8:9-12. September 1923; Palmer, Roy A. "Broader Economics of Safety." *The Industrial Engineer.* 83:363-365; 402; Mason, Robert C. "A Review," *Factory.* 38:96-97; 152-154, January 1927

[35] De Harts, Stanford. "Reducing Time Lost by Accidents," *American Machinist.* 64:967-969 June 1926

give attention not only to accident prevention but also to industrial hygiene and to the promotion of the general health of their workers by providing preventive and therapeutic medical care. In the larger establishments, plant hospitals and medical staffs were frequently provided and smaller employers found it advantageous to provide medical care for non-industrial disabilities through local medical facilities.

Trade Union Activity.—Intervention of government and efforts of employers to alleviate conditions were restricted largely to the more specific hazards of industrial accidents and occupational diseases. Simultaneously, trade unions were concerned with reduction of hours and increase of wage rates in order that the industrial worker might have, among other things, vitality equivalent to that of others. Largely through collective action of trade unions the length of the working day in America has been reduced from about 12 or more hours to around 8 hours.[36] It seems reasonable to suppose that lessening of fatigue and added hours for relaxation have exerted a favorable influence upon the wage earner's health.[37]

Technological Changes.—Another set of forces presumably having favorable implications for the health of the worker, lies in the evolution of the industrial system through the development of machinery that has lightened the physical exertion involved in a day's labor and enabled the semi-skilled worker to change occupations. Rapid development in facilities for human transportation enable workers to change jobs more frequently and to live further from factory districts. But the picture may not be one of clear gain, for while change has eliminated many

[36] See *Handbook of Labor Statistics.* U. S. Bureau of Labor Statistics Bulletin No. 616. Washington: Government Printing Office. 1936. P. 879. See also Beney, M. Ada. *Wages, Hours and Employment in the United States, 1914-1936.* New York: National Industrial Conference Board, Study No. 229. 1936.

[37] See Warren, B. S. "Industrial Conditions: Their Relation to the Public Health." *Public Health Reports.* May 29, 1914. Reprint No. 195. P. 6

health hazards it appears to have substituted some others. It is contended by labor leaders that increased monotony and tempo of work associated with mechanization has done much to offset the favorable effects of shorter hours.[38]

On the whole, by the beginning of the depression the combined forces of social and economic control had brought about considerable improvement in working conditions by (1) improving the standards of safety and hygiene and (2) shortening the working day. At the same time the unfavorable factors of monotonous mechanization of labor and production "speedup" had entered the industrial picture.

B. Working Conditions and the Depression

The degree to which each of the foregoing controls is exercised is probably influenced by economic conditions. Laws may remain on statute books but the will to enforce them is not immutable and even public opinion may countenance relaxation. In an effort to reduce costs, employers may curtail voluntary hygienic and medical services. The forces of organized labor tend to become disorganized during long periods of industrial depression unless they are bolstered by government support, with the result that other gains in the alleviation of working conditions may be temporarily lost also.

Standards of Safety and Hygiene.—Factors related to industrial safety and hygiene include the use of safety devices and machine guards, illumination, ventilation, protection against poisonous substances, plant housekeeping, provisions for adequate floor space and proper clearances, and maintenance of tools and equipment, vehicles, containers, electric insulation, and so forth in safe working condition. Others factors of importance are the lessening of mental hazards by provision for compensation, medical and hospital care, and training or instruction of workers against hazards.

[38] *Ibid.*

The use of safety devices and machine guards, standards of maintenance of boilers, pressure tanks, elevators, and so forth are governed in most states by safety codes and standards of safe practice.[39] In times of economic stress when state governments are faced with heavy deficits, it might easily occur that budgets for the maintenance of factory inspection and accident prevention enforcement were curtailed below the point where regulation can be considered effective. Likewise, employers faced with heavy operating losses may have relaxed voluntary safety and hygienic practices. They may have delayed renewing and renovating worn tools and equipment thereby increasing the hazards of accidents. Workmen's compensation laws which have contributed to the reduction of specific hazards of industry, may have been changed during the depression by state legislatures and modified by court decisions. A tendency to relax the economic burden imposed upon the employer would tend to encourage him to lower safety and sanitary standards.

Hours of Work and Speedup.—With the advent of the depression it appeared to some observers that a trend toward longer working hours was under way.[40] During the life of the National Industrial Recovery Administration, this trend was reversed, but following the invalidation of the Act the movement toward longer hours was resumed.[41] At the same time it was being charged by unionists that, in an effort to meet added costs, employers in textile mills were giving workers a larger number of looms to tend than was customary before the depression, that automobile assembly lines were being operated at a faster pace, and that similar methods of increasing the output of labor were being adopted elsewhere. These developments are important to the student of health for it has been seen that there is probably

[39] *Handbook of Labor Statistics.* 1936. Pp. 306-307

[40] Wolman, Leo and Peck, Gustav. "Labor Groups in the Social Structure." *Recent Social Trends in the United States.* New York: McGraw-Hill Co. 1933. II. Ch. XVI. Pp. 829-830

[41] *Handbook of Labor Statistics.* 1936. P. 879

a definite relationship between the frequency of industrial accidents and the length of the working day, fatigue, and nervous strain.

c. Studies Suggested

The proposed researches into the effect of the depression on occupational environment fall into two classes: (1) changes related to standards of safety and hygiene, hours of work, and physical demands made upon the worker by speedup technique and (2) studies of industrial accidents as related to use of safety devices, condition of machinery, fatigue, and selection according to length of time unemployed.

Environmental Studies

(1) Workmen's Compensation Laws.—Because workmen's compensation laws are important in securing improved standards of safety and hygiene and in relieving the economic burden of the worker and his family at the time of an accident, it will be of value to study changes in these laws throughout the depression years. The United States Bureau of Labor Statistics, Bulletin Number 496 contains a digest of all state compensation laws as of January 1, 1929.[42] Texts for later years may be had from the various state commissions. Items that should receive particular attention in such a study are: (1) changes in the number and proportion of workers and industries covered by compensation laws (2) changes in the compensability of occupational diseases and (3) changes in amount of compensation either by legislation or by decline in earnings used as a base in fixing the amount to be paid.

(2) Safety Codes and Their Enforcement.—Statutory provi-

[42] *Workmen's Compensation Legislation in the United States and Canada as of January 1, 1929.* U. S. Bureau of Labor Statistics, Bulletin No. 496. Washington: Government Printing Office. 1929

sions for use of machine guards and other safety devices and those establishing hygienic standards lend themselves to study similar to that suggested for compensation laws. In addition, data are available in several states for determining whether there were depression changes in the number of personnel or amount of expenditures for the enforcement of regulations, the number of citations for violation, and the amount of fines assessed. Changes should be calculated with reference to the number of plants in operation or the number of workers employed.

(3) Expenditures for Accident Prevention.—Expenditures for this purpose can be studied in some states for predepression, depression, and recovery years from employers' accounts which are submitted to compensation commissions for audit, or from data made directly available by employers.

(4) Hours of Work.—The Bureau of Labor Statistics has published data on weekly hours of work in a wide number of industries.[43] These data may prove sufficient to reveal the extent of changes in the length of the working week with sufficient accuracy to be useful for relating them to the problem of health.

Industrial Accidents

Studies of changes in the total number of accidents were suggested in Chapter II. Data are available for making detailed studies of accident rates in relation to certain environmental factors. In Michigan, for example, official records yield information relative to the nature of the industry; occupation, age, sex, marital status and the nativity of the employees; hour of occurrence of the accident and the time the injured employee began work on the day of injury, length of working day and the working week; cause of accident; cause and nature of injury; and the severity and degree of disability of the accident. The following studies are suggested:

---

[43] *Handbook of Labor Statistics* and *Monthly Labor Review*

(5) Safety Devices.—Trends in accident rates connected with the use of particular machines can be computed over a period of years and studied with reference to changes in legislation governing the use of safety devices, amount of inspection service, etc.

(6) Depreciation of Equipment.—Accidents are reported in terms of cause of injury, as broken arm due to the failure of a hoist, and in terms of cause of the accident, as failure of a hoist due to worn cable. The depression may have brought about reduction of expenditures for the upkeep or replacement of machinery and studies should therefore be made of trends in accidents resulting from mechanical failure.

(7) Fatigue.—Accident rates may also be studied with reference to changes in the number of hours in the working day and working week.

Selection of Workers

Studies of the health of selected communities made during the early years of the depression indicate that those who became unemployed soon after the depression set in have significantly higher rates of physical impairments and disabling illnesses than employed persons and the general population.[44] This is taken to indicate that unemployment during the depression is selective as to physical types—those individuals most physically fit retaining their employment status at the expense of the less robust workers. Data are available from two sets of sources to throw light on the subject of differential accident rates among different groups of employees.

(8) Public Works Administration Projects.—Contractors operating under this agency were required to use relief workers as far as they were available and then other unemployed persons. In general, the unemployed who were on relief had been out

[44] Diehl, Harold S. "Physical Condition and Unemployment." *Public Health Reports.* November 15, 1935. 50:1610-1618. No. 46

of work for longer periods of time than other unemployed and had more disabilities.[45] Accident rates can be computed for these workers after they have been classified according to occupation and relief status. Some projects like the Bonneville and the Coulee Dams employed several thousand workers for periods as long as three years. Since most of these men came within the provisions of state compensation laws, the records of state agencies can be regarded as a source of data for such a study.

(9) Works Progress Administration Projects.—Many projects conducted by this agency provided work in occupations identical with those of private industry. It seems that it should be possible, therefore, to compare accident rates on these projects with those in private industry to determine whether the rate among the emergency workers was higher than that among those who had not lost their regular jobs. Employees on work projects were usually covered by compensation laws and the Works Progress Administration kept detailed records of the number employed and the hours worked. Care should be taken to determine that those employed on work projects were following occupations to which they were accustomed, which was not always the case.

CHANGES IN SOCIAL ENVIRONMENT

The foregoing discussion has drawn attention to a group of economic factors which are presumed to affect the health of the population exposed to them. Brief consideration is now given to certain aspects of social organization and customs, such as family life, education, communication, recreation, social work, and population mobility, which are also presumed to be related to health even though the relationship has not yet been measured to any great extent. Since most of these factors are the subjects

[45] Palmer, Gladys L. and Wood, Katherine D. *Urban Workers on Relief.* Part 1. Works Progress Administration Research Monograph No. IV. Washington: Government Printing Office. 1936; preliminary unpublished data of the National Health Inventory

of companion memoranda, the present discussion is limited to a series of brief sketches indicating some of the possible implications for health. It is hoped that those who initiate researches into depression changes affecting these institutions will consider them in relation to health.

A. The Family

Family organization may influence health both directly and indirectly. The correlation between economic status and health suggests the influence which the decrease in family size may be having on the reduction of infant, child, and maternal mortality and the improvement of the health of the parents in the early adult and middle ages. Possible acceleration during the depression of the trend toward smaller families, particularly among the lower economic classes, would be significant in this connection. Deferred marriages, one of the most consistent depression phenomena, may have had quite the opposite effect on the emotional life of those prevented from marrying and upon the venereal disease rate. Likewise, delayed marriages may have given rise to an increase in abortions to unmarried women or, on the other hand, the high fee for producing an abortion may have resulted in allowing more pregnancies of unmarried women to proceed to full term and thus explain the increase which occurred in the illegitimacy rate.[46]

The criminal definition given to the practice of abortion makes it nearly impossible to obtain statistics of its incidence. The few studies that have been made[47] should be continued even though

[46] Stouffer, Samuel A. and Spencer, Lyle M. "Marriage and Divorce in Recent Years." *The Annals of the American Academy of Political and Social Science.* 188:56-69. November 1936

[47] See Stix, Regine K. "A Study of Pregnancy Wastage," Milbank Memorial Fund *Quarterly.* 13:347-365. No. 4. October 1935; Taussig, Fred J. "Abortion in Relation to Fetal and Maternal Welfare." *Fetal, Newborn and Maternal Morbidity and Mortality.* A publication of the White House Conference on Child Health and Protection. New York and London: D. Appleton-Century Co.

the possibility of establishing trends is remote. Data concerning the incidence of venereal diseases are also fragmentary. Nearly all states require physicians to report cases under their care[48] but even if these reports were complete, they would not include cases treated by quacks and drug stores and those receiving no treatment. It is to be hoped that the recent movement to remove the cultural taboos surrounding venereal diseases will yield better statistics as a byproduct.

The depression economy of doubling up among families may have resulted in psychological unbalancing and loosening of moral restraints which, in turn, may have resulted in decreased vitality. The *Research Memorandum on the Family in the Depression* by Samuel A. Stouffer and Paul F. Lazarsfeld, contains a further discussion of these questions.

B. Education and Communication[49]

Association between health status and education, outside of the field of health education itself, is difficult to establish. Nevertheless, a moment's speculation suffices to indicate the probable existence of a relationship. The effectiveness of a program of health education presupposes a cultural background broad enough to enable the members of the community to appreciate the personal and social significance of disease-preventing measures. The media through which the cultural base is broadened are largely formal and informal education, effected through the schools, the press, the radio, and motion pictures, all of which were doubtless influenced by the depression in ways that may

1933; Millar, William M. "Human Abortion." *Human Biology.* 6:271-307. No. 2, May 1934; Taussig, F. J. *Abortion, Spontaneous and Induced, Medical and Social Aspects.* St. Louis: C. V. Mosby. 1936

[48] Fowler, E. "Laws and Regulations Relating to Morbidity Reporting." *Public Health Reports.* Supplement No. 100. 1933. Pp. 4, 7

[49] See Educational Policies Commission: *Research Memorandum on Education in the Depression* and Waples, Douglas: *Research Memorandum on Social Aspects of Reading in the Depression.* (monographs in this series)

have affected health. In the field of formal education an increase in the enrollment in secondary schools and institutions of higher learning would tend to make the nation more receptive to the practice of measures leading to increased well-being.

Radio is utilized not only to promote recognized health agencies and practices but also to further the sale of innumerable medicinal preparations, some of which are of value, some of which are innocuous, and a great many of which are harmful. The long continued use of radio advertising by distributors of these preparations is eloquent of the value which this social development must have to them. It would be significant to know the extent of changes in the use of the radio for these purposes during the depression.

It is believed that the public press became more conscious of matters of health. The consequences of the depression among families contributed to more frank public discussions of the subject of providing medical care, not only for those on relief but, also, those who even in normal times have insufficient income to purchase medical and hospital care. The attitude of the press toward patent medicine advertisements may change with the state of prosperity of the publications. Publishers who normally restrict their columns to apparently bona fide preparations may attempt in lean times to bolster declining advertising space by accepting advertisements of unrecognized preparations. It would be significant, therefore, to know whether or not any general changes in this policy occurred during the depression, and it is suggested that a study of the relative number of inches of patent medicine advertising running in newspapers and periodicals before and during the depression should be incorporated in any researches which may be undertaken in this field.

c. Social Work Other Than Relief

The strengthening or weakening of those organizations devoted to social work other than direct relief must not be ignored in the present consideration. In the past the responsibility for the rehabilitation of dependent families, the care of dependent children and wayward girls, provision for recreation and healthful physical exercise among children of under-privileged families and homeless adults has fallen largely upon voluntary institutions engaged in social work.

A complete breakdown of these social facilities during a period of economic stress would conceivably result in an increased incidence of physical and mental impairments among those who customarily receive these services. It is very likely that public relief has adequately supplied, according to relief standards, the physical necessities of those formerly dependent for care upon private charities which may have had to suspend operation during the depression. It is not at all certain, however, that a curtailment of the services of organizations for private social work will not have resulted in deleterious mental reactions among a portion of the customary recipients of this type of relief. *The Research Memorandum on Social Work in the Depression* by F. Stuart Chapin and Stuart A. Queen contains discussions of the activities of medical social workers, psychiatric social workers, and nurses during the depression.

D. MOBILITY[50]

The population of the United States as individuals, households, or classes, is mobile from two standpoints: (1) changing residence, i.e., internal and international migration, and (2) carrying on the more ordinary activities of life—working, going to school, attending church, visiting friends—i.e., *fluidity.*

Whether a man moves from Bavaria to Iowa or merely walks from his home to the theatre, there is a change in the health

[50] See, in this series, Thompson, Warren S. *Research Memorandum on Internal Migration in the Depression.* Also: Young, Donald. *Research Memorandum on Minority Peoples in the Depression* and Sanderson, Dwight. *Research Memorandum on Rural Life in the Depression*

situation. He leaves a customary environment for one requiring adjustment. Since movement, especially that involving change of residence, is made to secure greater economic, social, and cultural advantages, the individual's mental health is probably influenced by the extent to which he believes the purpose of his migration has been realized. Physically he may have to adapt himself to strange people, different living and working conditions, unfamiliar standards of sanitation and quarantine and new diseases.

The movements of a person likewise have a bearing upon his surroundings. When he changes locations, the health situation becomes different for (1) the group he has left, (2) the people he encounters en route, and (3) the group he joins. In other words, the original environment loses certain factors of its equilibration, and the new one is the recipient of elements which may necessitate re-balance. What is more, the greater the number of individuals leaving or entering a situation, the greater the possibility of upsetting its health equilibrium.

Historically the movement of peoples has been associated with the spread of disease. This is illustrated from the earliest times by the spread of venereal diseases throughout the world and the introduction of smallpox among the American Indians through contacts with European invaders. A striking example of our own day is the development of the severe tropical forms of malaria in Russia shortly after the World War. At this time, tropical malaria which had been experienced only in the Caucasus and Turkestan, spread as far as Moscow and the vicinity of the Arctic Circle. The All-Russian Congress on Malaria held in Moscow, January 1923, ascribed the increase partly "to the mass movements of population (migration, repatriation, movement of troops)."[51] The strict quarantine regulations enforced

[51] *Epidemiological Intelligence.* Geneva: League of Nations. Health Section. No. 7, 55-59, 1922, and No. 8, 36-38, 1923

by the United States Public Health Service at every port are designed largely to prevent the spread of disease by incoming travelers and immigrants.

Theoretically any force which tends to slow down or inhibit the movement of peoples limits also the spread of contagious diseases and decreases the danger of epidemics. Whether mobility is limited by economic conditions or is brought about by administrative action, such as closing schools and prohibiting public assembly as a public health measure, makes little or no difference from the standpoint of community sanitation; except in degree, the effect upon health is the same.

There is evidence that the present depression greatly inhibited the mobility of our population. At any rate, this happened with respect to international migration, where the number of immigrants became so small that it was exceeded by more than 250,000 emigrants during the five-year period, April 1, 1930 to July 1, 1935.[52] Likewise checked were the internal movements of people from the farm to the city and from the city to the farm.[53] Recent researches have indicated that the popular idea that the depression was accompanied by a large back-to-the-farm-movement is erroneous.[54]

Although greater unemployment might lead one to expect more movement during a depression, this apparently was not true as regards interstate and intercounty migration and fluidity.[55] Limitation of income, the nationwide extent of the adversity, and the residential policy of relief giving presumably were the inhibiting factors. The strength of these factors in checking

[52] "Estimated Population of the United States." News Release. U. S. Bureau of the Census. February 4, 1936

[53] Dorn, Harold F. and Lorimer, Frank. "Migration, Reproduction, and Population Adjustment," *Annals of the American Academy of Political and Social Science.* 188:280-89, November 1936

[54] *Ibid.*

[55] *Ibid.*

intra-county moves, however, seems doubtful. Certainly there was much shifting about within the counties and cities of the United States in an effort to make a living; to save on rent, transportation, and other expenses; and to be near sources of relief.

Outside of the successive invasion of urban residential neighborhoods by different racial or cultural groups, little attention has been given to the purely local phases of mobility. An intensive study of these might reveal more of the exact relationship between mobility, health, and depression than a great deal of emphasis upon international migration or upon internal movements between regions and states.

Chapter IV

The Prevention and Treatment of Illness

ORGANIZATIONS and programs for conservation of health and care of illness are environmental factors which owe their existence to conscious efforts of individuals to protect the health of themselves and of others. Among institutions concerned with the prevention of illness are health departments with programs for communicable disease control, preschool and school health promotion, sanitation and food inspection, health education, prenatal care, and immunization; public schools; and private health agencies. Organizations for the care of illness include hospitals, out-patient departments, medical insurance plans, medical, dental, and nursing schools, and the practice of medicine and related professions. Although the effects of specific changes in public health programs or in the availability of medical care are difficult to measure in terms of the amount of sickness experienced, they are nonetheless important from the viewpoint of immediate and future comfort and health. It was said in Chapter I that possible changes in these institutions are not to be regarded as indexes of similar changes in the incidence of illness.

Extensive and prolonged drop in individual and community purchasing power may well have brought about changes along several lines among which are the following: (1) decline in family income undoubtedly implied change in the amount of preventive and therapeutic care that could be bought; (2) decline in hospital income from pay patients and from other sources may have reduced the number of patients cared for at less than cost rates; (3) reduction of appropriations may have restricted preventive activities of health agencies; (4) inability of large groups of the population to pay for medical care and of private agencies to provide care may have caused governments to step in with new funds and new methods for both prevention and treatment; (5) such an extension of government activity may easily have resulted in public demands that the government continue and expand its work along these lines; and (6) shortage of pay work for professional persons and for private agencies may have led to modification of the organization for the prevention and care of illness and to increased efforts to spread the costs of illness.

It is easily recognized that attempts to gauge the influence of the depression on the institutions and practices in the prevention and care of illness lead to studies along several widely diversified lines. The present chapter considers problems in three fields: (1) volume of preventive services (2) amount of medical care received and (3) organization for provision of preventive and therapeutic services.

THE VOLUME OF PREVENTIVE SERVICES

The health of a people is conditioned by an infinite variety of circumstances. Hence, measures of efforts to promote and preserve it are also numerous and varied.

A. Environmental Control

Community services designed to prevent sickness through control of the environment include the promotion of sanitation and cleanliness, safety, and disease control. Many of these services are identified below to indicate the scope of the field that must be covered in order to arrive at an evaluation of the total effects of the depression on health work.

Sanitation and cleanliness includes provision of an adequate supply of pure water; safe disposal of sewage; collection of garbage and refuse; building of sanitary privies; construction and maintenance of convenience stations, drinking fountains, swimming places and bath houses; street cleaning; smoke, dust, and noise abatement; sealing abandoned mines; swamp drainage and insect control; domestic animal control and eradication of rodents; promulgation of rules for and inspection of establishments preparing and handling food, milk, and other beverages; examination of plumbers and inspection of plumbing; inspection of buildings and factories for unhealthy conditions. *Promotion of safety* includes building and electrical inspection; supervision of working conditions; traffic control; inspection and regulation of elevators and fire escapes; and snow and ice removal. *Control of disease* involves diagnosis, reporting and isolation of communicable diseases; location of the sources of venereal, tuberculous and other infections; regulation of places of public assembly; promulgation of anti-spitting and similar ordinances; control of diseased animals and meat; examination of food-handlers; collection and analysis of records of births, deaths, and cases of infectious diseases.[1]

Multiplicity of Health Agencies

The number and type of agencies providing the services just listed are almost as numerous as the services themselves, although most of the work is done by city, county, and state health departments and the United States Public Health Service. Many

[1] Detailed statements of these activities are to be found in *Report of the Committee on Municipal Health Department Practice of the American Public Health Association* in cooperation with the United States Public Health Service. July 1923. Public Health Bulletin No. 136; *Municipal Health Department Practice for the Year 1923*, based upon surveys of the 100 largest cities in the United States made by the United States Public Health Service in cooperation with the Committee on Administrative Practice. American Public Health Association. Public Health Bulletin No. 164. July 1926; and *Public Health Organization: Report of the Committee on Public Health Organization. Section II, Public Health Service and Administration.* White House Conference on Child Health and Protection. New York: The Century Co. 1932. Attention is called to the fact that snow removal, sewage disposal, smoke and dust removal, and certain other activities are frequently not regarded as functions of health departments but of other government agencies. Discussion of them is included here because of their probable relation to physical and mental well-being rather than because they may or may not come within the province of any particular government department.

of these functions are performed by departments of public works, sanitation, water supply, sewage disposal, street cleaning, building and plumbing inspection, etc. Similarly, other federal agencies share responsibility for public health work with the Public Health Service. The Department of Agriculture supervises the preparation of animals for food, the interstate shipment of meat, food, and milk, and the distribution of drugs. It is also concerned with control and eradication of tuberculosis in cattle and swine, and of rats, ticks, and other rodents and insects; the improvement in strains of medicinal plants; investigation of farm water supplies; and drainage of surplus water. The Bureau of Internal Revenue is charged with enforcement of the narcotic laws. The Department of Labor and the Bureau of Mines develop standards for the control of the occupational environment of the worker. The Department of the Interior investigates water supplies, promotes flood control and, through the Office of Education, develops standards and programs for school hygiene. The Children's Bureau does research and sponsors programs for the improvement of maternal and child health. The Department of Commerce collects and tabulates vital statistics; does research on food, insect control, and food value of fish, and propagates fish; formulates safety codes and regulates air commerce and the use of electricity and gas; and registers chemicals, medicines, pharmaceutical preparations, medical appliances, etc.[2] During the depression the Works Progress Administration, the Public Works Administration, the Federal Emergency Relief Administration, the Rural Electrification Administration, the Federal Housing Administration, and several other agencies have promoted or engaged in many of the same activities designed to improve the health environment.

While control of the health environment is largely a governmental function, there are also a number of private agencies in

[2] Tobey, James A. *The National Government and Public Health.* Baltimore: The Johns Hopkins Press. 1926

the field. In many cities garbage and refuse disposal and the water supply are furnished by private concerns. Private research agencies and universities do much research on problems of public health, and industry develops most of its own safety devices.

Health Agencies and the Depression

Over the past few years, health agency programs have expanded as rapidly as health officers and others have been able to develop them and to persuade those who control government finances that they are worthwhile. The volume of expenditures and services fluctuates with government income, but the correlation between expenditures for health services and receipts from taxes is by no means perfect. Adjustment to reduced income may not be spread uniformly over all government departments and health services may suffer more or less than others. Some health services such as water supplies, baths and pools, and milk inspection are partially sustained by fees. To the extent that such charges cover costs, the provision of services becomes at least partially a function of the ability of individuals or groups to purchase them. During periods of severe depression both governments and private individuals are likely to restrict health activities. It is widely believed by health officers that this has been the case during the recent depression, and there are some published data to substantiate their beliefs. In the interpretation of these data, however, it must be borne in mind that, after 1933, emergency programs of the federal and state governments may have offset the declines occasioned by curtailment of regular appropriations.

Difficulties of Measurement.—The diversity of services and agencies gives rise to difficulties in measuring changes in activities. These difficulties, which must be considered in interpreting existing data as well as in connection with the suggestions for research, include:

(a) Marked variation in record systems from one community to another.

TABLE V

INDEXES OF EXPENDITURES FOR CONSERVATION OF HEALTH, BY TYPE OF SERVICE RENDERED: CITIES OVER 100,000 POPULATION, 1927–1934[a]

(AVERAGE 1927–29 = 100)

YEAR	CONSERVATION OF HEALTH								
	TOTAL	SUPERVISION	VITAL STATISTICS	PREVENTION AND TREATMENT OF COMMUNICABLE DISEASE		MEDICAL WORK FOR SCHOOL CHILDREN	OTHER CONSERVATION OF CHILD LIFE	FOOD REGULATIONS	OTHER CONSERVATIONS
				TUBERCULOSIS	ALL OTHER DISEASES				
1927	94	105	no data	94	96	92	95	95	no data
1928	99	95	98	98	97	102	98	101	82
1929	107	100	102	108	107	107	108	105	118
1930	113	82	107	126	102	128	106	120	195
1931	118	82	107	134	101	133	130	121	221
1932	107	78	93	117	87	128	121	111	225
1933	100	72	84	111	81	112	130	102	108
1934	108	67	92	121	90	111	153	115	123
Average yearly expenditures 1927–29 (in thousands of dollars)	118,231	14,775	1,531	36,477	26,553	12,929	9,658	7,973	188

[a] *Financial Statistics of Cities.* Washington: U. S. Bureau of the Census. Vols. for 1927–34.

(b) Lack of uniformity in the allocation of particular functions among various government departments.

(c) Measurement of services for the preservation of health in different terms such as volume of work done, expenditures, personnel employed, or population covered. One employee often performs a number of services and it is difficult to allocate his time or the cost to the different ones. No one of these units is suitable for measuring all types of services; it is impossible, for example, to estimate the number of people who profit from eradication of insects or rodents, although that is probably the most satisfactory method of measuring the work of an immunization program.

(d) Changes in purchasing power of the funds appropriated, which means that while expenditures dropped noticeably during the depression so also were wages reduced. Hence the volume of services probably did not decline to the same degree that expenditures dropped.

TABLE VI

INDEXES OF EXPENDITURES FOR PUBLIC SANITATION, BY TYPE OF SERVICE RENDERED: CITIES OVER 100,000 POPULATION, 1927–1934[a]

(Average 1927–29 = 100)

Year	Sanitation and Promotion of Cleanliness					
	Total	Sewers and Sewage Disposal	Street Cleaning	Other Refuse Collections and Disposals	Public Convenience Stations	Other Sanitation
1927	96	98	100	98	99	92
1928	100	98	97	102	100	105
1929	103	104	103	100	101	102
1930	106	96	79	149	99	98
1931	103	92	76	146	95	103
1932	97	88	61	153	90	102
1933	82	75	49	133	81	80
1934	76	65	53	113	77	78
Average yearly expenditures 1927–29 (in thousands of dollars)	364,384	66,746	171,532	263,453	2,329	13,904

[a] *Financial Statistics of Cities*, Washington: U. S. Bureau of the Census, vols. for 1927–34.

(e) Changing techniques, which may enable one employee or one dollar to perform twice the service formerly given, may cause expenditures or number of personnel to fail as indexes of trends in volume of work.

Expenditures and Changes in Urban Communities.—Since 1921, the United States Bureau of the Census has collected and published annually data on the costs of city governments with expenditures for health activities shown in considerable detail.[3] Until 1933 the reports covered all cities of 30,000 or more population; but in 1933 and 1934 they cover only cities of 100,000 or more population. Expenditures for cities of 100,000 or more population for the years 1927 through 1934 have been reduced to index numbers based on average expenditures for the years 1927-29, and are shown in tables V and VI.

[3] See *Financial Statistics of Cities*. Washington: U. S. Bureau of the Census. Vols. for 1927-34.

In the group of United States cities with populations of 100,000 or more in 1930 the index of expenditures for conservation of health of all municipal government departments increased from 94 in 1927 to 118 in 1931. In 1932 there was a sharp drop to an index of 107 which was followed by a further decline to 100 (the 1927-29 average) in 1933. Recovery to an index of 108 took place in 1934, perhaps as federal emergency funds became available to cities for these or other services. The movements of expenditures among the various subdivisions of the function of conservation of health are similar to that for all expenditures; in some cases the peak year is reached before 1931 and in some cases, later. In every instance, save one, however, the index number for the year 1933 is lower than for the years 1930 and 1931 and there is only one instance in which the expenditures for 1933 exceeded those for 1932. A similar picture is presented by the data for sanitation and the promotion of cleanliness, but the drops are more marked than among the activities included in conservation of health.

Per capita expenditures necessarily fluctuated in about the same way for these cities of 100,000 or more population. Expenditures for all conservation of health increased from $1.13 per inhabitant in 1927 to $1.32 in 1931. The year 1932 witnessed a drop to $1.11 and 1933 to $1.05. There was a recovery to $1.13 in 1934. For sanitation or the promotion of cleanliness per capita expenditures decreased from about $3.75 for the years 1927-30 to $3.13 in 1932, to $2.66 in 1933 and still further to $2.44 in 1934.[4, 5] It is to be remembered that these data are based on absolute dollar expenditures and that the purchasing power of the dollar for health and sanitation services presumably varied considerably during this period. Increased expenditures during the early stages of the depression probably resulted in even

[4] *Ibid.*

[5] Intercensal population changes may have affected these averages somewhat.

larger increases in services; on the other hand, decreased expenditures during the recovery period probably resulted in even larger decreases in services.

On the basis of these census data it appears that budget reductions were not spread evenly over all governmental activities. In 1927 conservation of health accounted for 2.8 per cent of the expenditures among these cities and sanitation or the promotion of cleanliness for 8.1 per cent. The declines from these

TABLE VII

EXPENDITURES FOR CONSERVATION OF HEALTH, AND FOR SANITATION AND THE PROMOTION OF CLEANLINESS AS A PROPORTION OF TOTAL EXPENDITURES: CITIES OVER 100,000 POPULATION, 1927–1934[a]

Year	Conservation of Health	Sanitation and Promotion of Cleanliness
1927	2.8%	8.1%
1928	2.6	7.5
1929	2.6	7.8
1930	2.6	7.4
1931	2.6	6.8
1932	2.3	6.5
1933	2.3	5.8
1934	2.4	5.3

[a] *Financial Statistics of Cities*, Washington: U. S. Bureau of the Census, Vols. for 1927–1934

proportions, which in part may reflect increased emergency expenditures in other fields, are shown in Table VII.

Budgetary declines during the depression are revealed in two other analyses of similar data. In a study made by the Federation of Social Agencies of Pittsburgh and Allegheny County the ratios of 1932 and 1933 expenditures to the 1929-31 average were computed. These are shown in Table VIII.

The data collected in connection with the Health Conservation Contests[6] show for 79 identical cities reported expenditures

[6] Health Conservation Contests among cities were developed in 1929 by the American Public Health Association and the United States Chamber of Commerce to encourage the standardization and development of public health programs. Rural counties were admitted to the contest in 1934. Contestants

TABLE VIII

CHANGES IN EXPENDITURES FOR PUBLIC HEALTH AND SANITATION, BY TYPE OF SERVICE: PITTSBURGH, 1929–31, 1932, 1933[a]

Service	1929–31 Average	1932	1933
Tuberculosis control	100[b]	91	91
Venereal disease control	100[b]	145	145
Other communicable disease control	100	77	65
Conservation of child life	100	88	72
Food and milk control	100	88	73
Laboratory service	100	87	99
Sanitation	100	67	49
Smoke regulation	100	77	59
Plumbing inspection	100	77	61
Garbage and refuse collection and disposal	100	89	68
All health expenditures	100	86	69

[a] Calculated from Ewalt, Marian H. and Hiscock, Ira V. *The Appraisal of Public Health Activities in Pittsburgh, Pennsylvania, 1930 and 1933.* Social Research Monograph No. 2. Pittsburgh: Federation of Social Agencies of Pittsburgh and Allegheny County. P. 8

[b] Average for 1930–31

for selected services[7] for the years 1930 through 1934.[8] Among these cities, expenditures increased from 86 cents per capita in 1930 to 90 cents in 1931 and then dropped to 83 cents in 1932 and to 73 cents in 1933. A rise to 77 cents followed in 1934. Per capita expenditures are given for each of the 79 cities as well as

---

report annually on expenditures and personnel, classified by functions, to the American Public Health Association. Ratings are given according to predetermined standards and plaques are awarded to winning communities by the Chamber of Commerce as described in "Health Conservation Contests," *Year Book,* 1935-1936, American Public Health Association. Although 1936 data have not yet been analyzed 255 cities and 193 counties participated during that year (see *News Letter,* Jan. 15, 1937. American Public Health Association). As will be suggested in the appropriate place, these records for individual cities provide data for significant studies of depression changes.

[7] The services include administration; vital statistics; general sanitation; milk, meat and other food control; acute communicable disease, tuberculosis and venereal disease control; child hygiene including maternal, infant, preschool and school hygiene; public health nursing and laboratory services.

[8] "Expenditures in Certain Cities for Selected Health Services," *American Journal of Public Health.* Vol. 26. No. 10. October 1936

for a large number of other places that did not report throughout the entire five-year period.

The study of the Pittsburgh Health Department provides a few statistics on personnel. In 1929 there were 357 employees in the department and by 1931 the number had increased to 423. But in 1933 the number had dropped to 391 or to a level 8 per cent below the peak. When hospital employees are excluded and consideration limited to those engaged in health conservation, the drop is 14 per cent between the two years.[9]

Changes in Rural Communities.—The number of rural counties having the services of a whole-time health officer is the subject of an annual report by the United States Public Health Service.[10] There are about 2,500 rural counties in the country. In 1931 some 557[11] of these counties had whole-time health officers and at the beginning of 1932 the number reached 616. During the year 1932 the number declined to 581. This was the first year since 1908, when the original rural county health department was established, that the number of departments discontinued exceeded the number established.[12] During the year 1933 further decline brought the total to 530, but the trend reversed in 1934 and by the end of 1935 there were again more than 600 counties with health programs directed by a full-time worker. In 1935, the 612 counties included 28.7 per cent of the rural population of the country.[13] Another study shows the year by

[9] Ewalt and Hiscock. *Op. cit.* P. 7. (See Table VIII, note a)

[10] See "Extent of Rural Health Service in the United States, Dec. 31, 1931 to Dec. 31, 1935." *Public Health Reports.* August 14, 1936; "Extent of Rural Health Service in the United States, Dec. 31, 1930 to Dec. 31, 1934." *Public Health Reports.* November 1, 1935

[11] *Ibid.* November 1, 1935. P. 1553

[12] See Mountin, J. W., Pennel, E. H. and Flook, E. E. *Experience of the Health Department in 811 Counties, 1908-1934.* Public Health Bulletin No. 230. Washington: United States Public Health Service. 1936. P. 7

[13] "Extent of Rural Health Service in the United States, Dec. 31, 1931 to Dec. 31, 1935." *Public Health Reports.* August 14, 1936. P. 1132

year fluctuations in the proportion of the population of states, exclusive of that in cities over 100,000, served by full-time county health officers; the proportion dropped slightly from 34.9 per cent in 1932 to 33.9 per cent in 1933.[14]

The full significance of these changes in the number of departments cannot be derived without reference to data showing the impact of budget reductions on the number of professional workers and the amount of service rendered. There has been no analysis of changes in volume of service, but a recent report[15] shows that discontinuation of some health departments and release of personnel from others reduced the number of medical officers by 6 per cent between 1931 and 1933; the number of sanitation officers by 12 per cent between 1932 and 1933; and the number of nurses by 15 per cent between 1931 and 1933. Data from another source reveal that between 1931 and 1935 there was a decrease of 5.4 per cent in the number of public health nurses regularly employed.[16]

Emergency Control Programs.—Thus far it appears that efforts to conserve health must have declined during the depression although this decline lagged behind the business index. While it is impossible at this time to undertake precise measures of these changes and to discover the net declines when counter changes are observed, there does seem to be some material that can be entered on the other side of the ledger. The Domestic Quarantine Division of the United States Public Health Service undertook four work projects in connection with the Civil Works Administration and Works Progress Administration programs: malaria control, a typhus fever project, community sanitation

[14] See Ferrell, John A. and Mead, Pauline A. *History of County Health Organizations in the United States, 1908-33.* Public Health Bulletin No. 222. Washington: United States Public Health Service. 1936. P. 45

[15] Mountin, Pennel and Flook. *Op. cit.* P. 16

[16] See "Tabulation of Health Department Services." *Public Health Reports.* September 4, 1936. P. 1347

(privy building), and sealing abandoned coal mines.[17] These and other Works Progress Administration projects accounted for considerable volumes of work. Through September 1936, more than 3,300 miles of sewers with 130,000 service connections had been constructed, more than 1,200 sewage treatment plants had been built or improved, 54,000,000 feet of drainage ditches had been dug, and over 3,000,000 feet of pipe laid.[18] More than 871,000 privies had been built in 41 states by June 1936,[19] and 43,000 coal mine openings had been sealed.[20] In addition, water mains, pumping stations, storage dams, water purification plants, and wells were constructed new or were improved at a cost combined with the above items of nearly half a billion of dollars,[21] and doubtless offset many of the regular government budgetary declines.

B. Health Education

Health education includes the dissemination of facts concerning the prevention of disease, the promotion of physical and mental well-being, and the purposes and operation of community health agencies including the methods of using them. Health education takes many forms and is provided on both mass and individual bases.[22]

Health Education: A Common Responsibility

Health education is one of the primary prerogatives of all departments of health from the smallest city or county unit up

[17] See Waller, C. E. "A Review of the Federal Civil Works Projects of the Public Health Service." *Public Health Reports*. August 17, 1934. Reprint No. 1642

[18] From unpublished records of the Works Progress Administration

[19] Letter to the authors from Dr. C. E. Waller, Assistant Surgeon General, Domestic Quarantine Division, U. S. Public Health Service, July 23, 1936

[20] From unpublished records of the Works Progress Administration

[21] *Ibid.*

[22] *Report of the Committee on Municipal Health Department Practice of the American Public Health Association.* (See note 1, p. 115)

to the United States Public Health Service. However, many other agencies participate. A considerable part of the funds for health work by the regular federal agencies is devoted to this purpose, particularly those of the Children's Bureau, the Women's Bureau, the Bureaus of Home Economics, Animal Husbandry, Agricultural Economics, Chemistry, Labor Statistics, Mines, Standards and the Office of Education. Likewise, during recent years, the emergency administrations, notably the Federal Emergency Relief Administration and the Works Progress Administration, have initiated or sponsored many projects in the field of health education. Public school systems and private organizations assume a large share of the responsibility for education. Among the latter are associations of visiting nurses, the American Public Health Association, the American Medical Association, the American Dental Association, the American Society for Control of Cancer, the National Education Association, the National Tuberculosis Association, the American Red Cross, the American Social Hygiene Association, the National Organization for Public Health Nursing, and the American Heart Association and many other state and local organizations affiliated with these national bodies. Similarly there are numerous other organizations with health programs. These include the Y W C A and Y M C A, civic organizations, women's clubs, Parent-Teacher associations, and settlement houses.

Increased Need for Education

It is almost axiomatic that the depression should have brought greater need for health education. Hundreds of thousands of families must have suffered income losses that made it difficult for them to purchase adequate supplies of food, clothing, housing, and medical care. It was important therefore that housewives and parents should be taught the most effective uses for the commodities they were able to purchase from their reduced incomes.

Funds for Education

Regular health agency appropriations declined during the depression. The data are not sufficiently refined to reveal whether these reductions reached programs for health education, but it is reasonable to suppose that they did. Private agencies, that might have assumed a greater proportion of the responsibility, dependent upon voluntary contributions, almost surely found sharp reductions in their own allowances. On the other hand, the depression may have stimulated interest in health education. Federal Emergency Relief Administration and Works Progress Administration contributions for public health nursing and home demonstration projects perhaps exceeded the probable decline of local health agency expenditures for those types of education; 185,000 health lectures and demonstrations had been held in 25 states by September 1936.[23] The distribution of food budgets and instructions for food preparation may have been increased as depression measures. The Social Security Act provided new funds for health education. These are being spent for maternal and child welfare education, general public health work, and industrial hygiene. Although there are more than 500 poisonous materials and hazardous conditions surrounding American workmen, only five State Health Departments had industrial hygiene programs prior to 1936. By November of that year, however, twelve other states had initiated programs and annual expenditures rose from $31,000 to $348,000. The new programs are establishing standards of personnel and cost, training workers, and making field studies.[24]

C. DIRECT SERVICES

Direct services for the prevention of illness are rendered primarily to individuals and include vaccinations and immuniza-

[23] From unpublished records of the Works Progress Administration

[24] See Sayers, R. R. and Bloomfield, J. J. "Industrial Hygiene Activities in the United States," *American Journal of Public Health.* 26:1087-96, No. 11, November 1936

tions, physical and dental examinations, prenatal and infant care, tuberculosis and venereal and other contagious disease case finding, and other services.[25] It has been stated earlier that many of these so-called direct services overlap health education and activities in the field of environmental control. There is considerable overlapping between these preventive measures and the treatment of disease discussed in a later section.

Public and Private Services

Direct service for the prevention of illness is given by physicians in private practice, public health departments and clinics, public schools, and private agencies including clinics, settlement houses, social hygiene groups, and such organizations as the National Tuberculosis Association. The proportion of services received from public clinics and private practitioners varies with the type of service. Studies undertaken by the Committee on the Costs of Medical Care and the United States Public Health Service indicate that 55 per cent of physical examinations and 42 per cent of all immunizations are made in public clinics, but only 16 per cent of prenatal care and post-partum examinations, 7 per cent of dental care and 3 per cent of all eye refractions are made by public clinics including those supported by private funds.[26]

Direct Services and Economic Conditions

Preventive and corrective services have been growing rapidly and in view of their potentialities for the conservation of health, it is significant to know whether the predepression trend was changed.

Changes in Services Rendered.—Earlier review of the ex-

---

[25] *Report of the Committee on Municipal Health Department Practice of the American Public Health Association.* (See note 1, p. 115)

[26] See Collins, Selwyn D. "Frequency of Health Examinations in 9,000 Families, Based on Nationwide Periodic Canvasses 1928-1931." *Public Health Reports.* March 9, 1934. P. 345

penditures of large cities for health functions indicated that appropriations for direct services declined during the depression along with those classified as environmental controls. Reference to Table V, however, reveals that, in general, prevention and treatment of communicable disease, medical work for school children, and other conservation of child life, declined less markedly and reacted more rapidly toward the end of the depression than did those for other functions. That there were reductions was substantiated by the data of the Health Conservation Contests and the Pittsburgh study.

Although smallpox vaccinations are compulsory, school medical inspectors in Pittsburgh examined only 34,000 children for vaccination in 1933 compared with 49,000 in 1930 and vaccinated only 8,800 in the latter year compared with 10,200 in 1930.[27]

TABLE IX

INDEXES OF NUMBERS OF VISITS OF PATIENTS TO SELECTED PRIVATE AND PUBLIC CLINICS: SELECTED URBAN AREAS, 1929–1935[a]

(1929 = 100)

Year	Index of Visits to	
	Private Agencies	Public Agencies
1929	100	100
1930	116	126
1931	134	158
1932	146	184
1933	148	207
1934	138	176
1935	139	180

[a] *Social Statistics Bulletin.* Children's Bureau, U. S. Department of Labor. No. 1, IV:2, 3. May 1936

The *Social Statistics Bulletin* of the United States Children's Bureau adds some data on trends during the depression.[28] Since 1929, the annual number of visits by patients to clinics and

[27] Ewalt and Hiscock. *Op. cit.* P. 31. (See Table VIII, note a)

[28] *Social Statistics Bulletin.* Children's Bureau, U. S. Department of Labor. Vol. 4:2, 3. No. 1. May 1936

health conferences has been reported by 108 private agencies in 16 urban areas and by 37 public agencies in 14 urban areas. (Table IX.) Data are also given for the annual number of visits by public health nurses on the staff of 16 public and 25 private agencies. Index numbers of the fluctuations over the same period are presented in Table X.

TABLE X

INDEXES OF NUMBERS OF VISITS OF PUBLIC HEALTH NURSES FROM SELECTED PRIVATE AND PUBLIC AGENCIES: SELECTED AREAS, 1929–1935[a]

(1929 = 100)

Year	Index of Visits by Public Health Nurses from	
	Private Agencies	Public Agencies
1929	100	100
1930	105	103
1931	109	104
1932	105	100
1933	96	84
1934	103	85
1935	107	87

[a] *Social Statistics Bulletin.* Children's Bureau, U. S. Department of Labor. 4:2,3. No. 1, May 1936

In both visits by patients and by nurses increases occurred in the first years subsequent to 1929, but they were followed by decreases during the middle depression period. It is also significant that in each of the four series of data some of the depression losses have been regained. Whether these recoveries represent increased expenditures from regular channels or the effect of emergency funds that found their way into regular government agencies is not known. As early as 1933 the United States Children's Bureau cooperated with the Federal Emergency Relief Administration and the Civil Works Administration in conducting child health conservation programs through the employment of 2,000 nurses, distribution of medical examination forms, and consultation of medical specialists with state and local relief, medical, and health officials.[29] More recently the Works Progress Administration employed nearly 7,500 public health and home nurses who made more than 3,500,000 home visits in the interest of disease prevention. In addition there were nearly 6,000,000 examinations of eyes, ears, teeth, and general health; 450,000 special tests including Schick, Wassermann, Dick, and Mantoux; 1,350,000 immunizations; and over 200,000 corrections of eyes, teeth, nose and throat and other disorders. More than 72,000,000 hot lunches had been served up to September 1936.[30]

Several of the direct services are furnished principally or in lesser part by private physicians. This means that the data cited above are deficient to the extent that they fail to cover the total volume of services; neither direct preventive services nor treatment for disease is spread uniformly over all elements of the population. Individuals in families which suffered income losses sufficiently great to thrust them near to or below the subsistence level must have suffered greater losses in preventive services than did those persons in families whose incomes did not decline to the poverty level.

Preventive Services and Family Income.—No data are available to show differential rates of reduction of preventive services among various groups but there are some that reveal the existence of differentials at given moments of time.

An analysis of data collected by the Committee on the Costs of Medical Care showed that the rate of eye refractions among persons in families with annual incomes of less than $1,200 was 22 per 1,000 population and that the rates increased with income

[29] See *Annual Report of the Secretary of Labor, 1934.* U. S. Department of Labor. Washington: Government Printing Office. P. 80

[30] "Works Progress Administration Health Projects Give Employment to 7,444 Women." Press Release No. 4-1275. August 16, 1936. Washington: Works Progress Administration; unpublished records of the Works Progress Administration

to 102 refractions per 1,000 population in the income group of $5,000 and over.[31] Health examinations show a relationship similar to that of eye refractions.[32] Professor Paul A. Dodd, in a study made during the depression for the California Medical Society discovered that only 80 per cent of the illnesses experienced by members of families with annual incomes of less than $500 had been diagnosed by a physician against 95 per cent in families of $3,000 or more.[33]

Prenatal and postnatal maternity care should include 20 visits to a physician according to standards presented in the reports of the Committee on the Costs of Medical Care. The family surveys conducted by the Committee, however, showed that in the lowest income groups an average of only seven such calls per case were made and that the number ranged upward to an average of 13 calls in the highest income group.[34]

Immunizations, as reported by the Committee, were performed at the rate of 69 per 1,000 population in the income groups under $1,200, 49 in the $1,200-$1,999 group and thence upward to 120 per 1,000 in families with incomes of $10,000 or more.[35] The high rate for the lowest economic group is explained through the provision of this service through public channels. Further information on receipt of immunizations according to family economic status will be available from the analyses of

[31] Collins, Selwyn D. "Frequency of Eye Refractions in 9,000 Families Based on Nationwide Periodic Canvasses 1928-1931." *Public Health Reports.* June 1 1934. P. 8

[32] Collins, Selwyn D. "Frequency of Health Examinations in 9,000 Families." Pp. 329-331. (See note 26, p. 128)

[33] Dodd, Paul A. *Economic Aspects of the Practice of Medicine in California* (MS.) A report submitted by the staff of the California Medical Economic Survey to the Committee of Five for the Study of Medical Care of the California Medical Association.

[34] Falk, I. S., Klem, M. C., and Sinai, N. *The Incidence of Illness and the Receipt and Costs of Medical Care among Representative Family Groups.* Publication No. 26 of the Committee on the Costs of Medical Care. Chicago: University of Chicago Press. 1933. P. 136

[35] *Ibid.* P. 275

the Communicable Disease Study and the Health Survey both of which are current projects of the United States Public Health Service National Health Inventory. From the Communicable Disease Study it will probably be possible to discover whether there have been variations in vaccination and immunization rates over the past 10 or 15 years.

D. SUMMARY

As in all research a number of obstacles lie between the present state of relative ignorance and the ultimate objective of clearer knowledge. The large number of preventive services and the variation in methods of work of numerous health agencies present a complex and forbidding picture because of variations in definition of services, in allocation of functions among governmental departments, and because of overlapping among publicly and privately supported units and practitioners. This same multiplicity of functions and organizations with their attendant variation constitutes a serious limitation to the preparation and collection of statistics of expenditures, of units of work accomplished, or personnel employed classified in significant detail according to types of service rendered. A limitation of much of the data currently available is that they are given in terms of expenditures. Statistics of expenditures tend to lose their significance over time unless some means is at hand for interpreting them in terms of real purchasing power. It would be difficult to interpret the activities of health agencies in these terms over a wide area for expenditures are made chiefly in the form of wages and wage standards and wage fluctuations vary considerably from one part of the country to another.

While these difficulties are real no competent research worker need permit them to discourage him. The seriousness of the difficulties would probably increase as the area of coverage increased, but large scale studies are seldom necessary. Accurate and detailed analysis of depression influences on a single community or on one activity in the field of health conservation is likely to be more useful than superficial attempts to estimate changes on a broad scale.

Not many data are already collected and available for the student. However, there are vast quantities of raw material in the reports and records of both public and private agencies. These data may not be comparable among different places, but comparability will probably be found over a period of time in one or a small number of places.

There is great need for the development of comparable record keeping and reporting. The promotion and preservation of health involves the expenditure of tremendous sums of public and private money. Likewise, it is a field in which activities and expenditures are increasing rapidly. Hence record keeping with respect to activities, costs, and other factors is of the utmost importance to the community. Data are needed for city to city comparisons; for measuring changes in volume and standards of work; and, most important of all, for evaluating preventive measures in terms of populations reached and specific illnesses prevented.

E. Studies Suggested

(1) Collection of Statistics of Public Health Work.—An immediate need in the field of public health is for the development of forms and systems for recording and reporting statistics of activities, personnel, and expenditures and for the designation of a centralized agency to collect, tabulate, and disseminate these data. Health agencies render a wide variety of services at tremendous cost to the community. Statistics should be available for local health officers, for state, federal, and other subsidizing agencies, for research, and for the information of the public supporting the activities.

Accordingly it appears that the time is propitious for the investment of some agency, presumably the United States Public

Health Service, with the responsibility for collecting and summarizing the data that would become available.[36] The basis for such a program is even now being prepared by the conference of State and Territorial Health Officers with the collaboration of the United States Public Health Service and the United States Children's Bureau. These groups have developed a detailed form and instructions for the use of health departments in making periodic records of the volume of work done.[37, 38] Drs. K. Stouman and I. S. Falk have also developed a plan covering the type of data that should be collected and a classification that would be useful in reporting.[39]

(2) Trends in Public Health Work from Health Conserva-

---

[36] An abstract of the Report of the Sub-Committee on Current Practices of Health Departments and of the Committee on Administrative Practice of the American Public Health Association recently had this to say with reference to reporting of health department work: "Strange as it may seem, there has never been developed in this country any comprehensive national system for reporting data of special interest to health officials, other than morbidity and mortality. . . . It is accepted as a principle by the Committee on Administrative Practice that a comprehensive system of reporting to be effective must eventually become a function of an agency in the federal government which is equipped for collecting data of this type and making the findings available." See Mountin, J. W. "A Central Information Service on Current Practices of Health Departments." *American Journal of Public Health.* 25:347. No. 3. March 1935

[37] "Tabulation of Health Department Services." *Public Health Reports.* September 4 1936. Pp. 1236-1251

[38] At the present time the suggested report form makes no provision for recording statistics of expenditures or of personnel employed. There is no provision for submitting these reports to a central agency for analysis. It is the experience of the American Public Health Association that local health officers request information concerning personnel and expenditures of other cities more frequently than they ask for other types of data. Mountin. *Op. cit.* P. 347

[39] Stouman, K. and Falk, I. S. *Health Indices: A Study of Objective Indices of Health in Relation to Environment and Sanitation. Quarterly Bulletin of the Health Organization.* Geneva: The League of Nations. Vol. V. December 1936. P. 901-1081. This excellent report contains a detailed schedule designed for use in connection with the suggestions and also a section describing the health indexes of the city of New Haven, Conn.

tion Contest Data—Personnel and Expenditures.—The data available from the forms submitted to the Health Conservation Contest committee are well worth analysis from the point of view of our subject matter. The forms have been expanded from year to year so that data in the present detail are not available from the inception of the contests in 1929 to date. Nevertheless, there are statistics of expenditures and personnel classified by type of service. Hence it is possible to determine trends among the various services for cities in different geographic regions and in different size groups, provided it appears that the cities making reports are not the best cities and hence unrepresentative of all cities in the matter of trends.

(3) Health Conservation Contest Data as a Basis for Study of Future Trends in Volume of Preventive Services.—The 1936 Health Conservation Contest form for use by cities provides for reporting the actual quantities of work done in connection with many specific types of health services.[40] Since future changes in these forms will probably retain most of the questions now asked, it is suggested that 1936 be established as a base year from which to measure changes in these services as economic recovery advances.

The American Public Health Association has established standards against which to measure the performance of local health departments in connection with determining the ratings for the Health Conservation Contests. The significance of the effects of the depression might appear more realistically if studies such as those just mentioned were made in terms of variation from these standards.

(4) Studies of Preventive Work from Records of Local Agencies.—The studies suggested thus far will have shortcomings

[40] These services are listed on the Health Conservation Contest forms which may be obtained from the Secretary of the American Public Health Association or from the United States Chamber of Commerce.

because of incompleteness of records and because a relatively small number of cities are included. Local health agencies and other organizations will want to know what happened to their programs in comparison with those of other cities during the depression. Therefore, it is suggested that studies limited to even one city will find interest if the analysis covers all significant services in proper detail. In a great many places health department records will yield data year by year on the volume of work performed, expenditures, and number of personnel employed.

(5) Preventive Work of Private Health Agencies.—Similar studies may be made of programs of private agencies in the health prevention field. Studies of private agencies should be combined with the analysis of data for public agencies in order to arrive at a total picture of depression effects.

(6) The National Health Inventory as a Basis for Future Studies of Trends.—The Health Facilities Study, a National Health Inventory project, has determined the expenditures, the volume of service given, and the number of persons employed during the year 1935 for all public and private agencies doing any type of health promotion work in each of approximately 100 urban counties scattered over the United States. The data are being analyzed according to a detailed classification of services. While this study covers only a fixed point in time it will provide a base against which future inventories may be compared.

The data collected in connection with the National Health Inventory should be particularly complete because rather extended effort was made to locate all agencies and because each agency was visited by a member of the field staff.

(7) Extension of Federal Health Services.—The regular activities of federal government agencies can be measured in terms of volume of work accomplished and expenditures. Particular

attention should be given to the extension of existing health functions and the development of new activities in response to the availability of emergency funds. Significant studies can be made of individual services, although an attempt to estimate the changes in the entire federal program would constitute a more important contribution.

(8) Health Work in Emergency Programs.—The development of health promotion activities in connection with the emergency relief programs were numerous and varied. The whole field should be studied and evaluated to gain an understanding of the potentialities of the supplementing of ordinary health work in periods of depression and for the use of these activities as satisfactory work projects for the unemployed.

(9) Depression Trends in Health Promotion Measured by Financial Statistics of Cities.—Reduced municipal expenditures for public health were not spread uniformly over activities for which data are available in the *Financial Statistics of Cities.* The data of these reports should be analyzed further to discover the extent of change in the health program. This study can be undertaken by geographic regions and by cities classified according to size. In connection with these studies it must be remembered that the purchasing power of the dollar changed, and that since the largest single item of expenditure in health work is for salaries and wages of government employees, the real purchasing power of the health promotion dollar must be measured in those terms.

(10) Trends in Health Services in Rural Counties.—There were marked changes during the depression and the early recovery period in the number of rural counties having whole-time health officers. An extension of this study would undertake to determine what factors beyond the shortage of funds were responsible for the abandonment of health programs in some counties and the initiation of work in others at various stages

in the business cycle.[41] Much of this analysis has been done and detailed data are available in a report by Ferrell and Mead of the Rockefeller Foundation.[42]

(11) Changes in School Health Programs.—Records of school health programs provide a useful source of data for studies in that field. Changes should be noted in the proportions of all entering and all other pupils given physical examinations; the extent to which home visits are made to secure correction of physical and dental defects; and other measures of specific health activities. Variations in the number of physicians, dentists, nurses, special teachers, and other persons employed on health work should also be studied.

(12) Health Topics in Newspapers and Periodicals.—It has been said that newspapers and periodicals constitute one of the most important media through which health education is carried on. A rather simple study can be made of the number of column inches of health news and the proportion that health news constitutes of all news in the papers of one or more com-

[41] See Mountin, Pennell and Flook. *Op. cit.* (Note 12, p. 123); and "Extent of Rural Health Service in the United States, December 31"

1935	Public	Health	Reports	Aug. 14, 1936	Reprint	No.	1764
1934	"	"	"	Nov. 1, 1935	"	"	1714
1933	"	"	"	Dec. 7, 1934	"	"	1661
1932	"	"	"	Oct. 6, 1933	"	"	1597
1931	"	"	"	Dec. 16, 1932			
1930	"	"	"	Sept. 11, 1931	"	"	1509
1929	"	"	"	May 9, 1930	"	"	1372
1928	"	"	"	May 17, 1929	"	"	1284
1927	"	"	"	Apr. 13, 1928	"	"	1220
1926	"	"	"	Apr. 29, 1927	"	"	1155
1925	"	"	"	May 7, 1926	"	"	1079
1924	"	"	"	May 8, 1925	"	"	1010
1923	"	"	"	May 16, 1924	"	"	921
1922	"	"	"	Apr. 27, 1923	"	"	833

[42] Ferrell and Mead. *Op. cit.* (Note 14, p. 124). See also monograph in this series by Sanderson, Dwight: *Research Memorandum on Rural Life in the Depression.*

munities over a period of years. Similarly, changes in the frequency of health topics in magazine articles can be discovered by using the *Readers Guide to Periodical Literature* in the manner employed by Professor Hornell Hart.[43]

(13) Trends in Immunization Work.—Some health departments that do few immunizations, none the less receive reports of immunization done by private physicians. Where reporting is satisfactory it is possible to establish year by year trends in the volume of this work throughout the period of prosperity and depression. The study should determine separately the trends of free or clinic immunizations and of those done in private practice.

(14) Proportion of Children Immunized Over a 19-Year Period from National Health Inventory Data.—The Communicable Disease Study, which is a part of the National Health Inventory, obtained from 214,000 families a record of the ages at which all children born to the head of the family since March 1, 1911 were vaccinated for smallpox or immunized for typhoid, diphtheria, or scarlet fever, together with age at the time of the canvass. These data provide the basis for a study of the trend of vaccinations and immunizations among children under 5 years of age for the period 1916 through 1935. The 1935 income and the occupation of the father are recorded for each family so that fluctuations in vaccination and immunization rates can be correlated with these items. Since trends will probably vary with the type of immunization program carried on by the health department, this factor too should be considered.

THE RECEIPT OF MEDICAL CARE

Care for the treatment of sickness implies a variety of services rendered by institutions and professionally trained persons

[43] Hart, Hornell. "Changing Social Attitudes and Interests." *Recent Social Trends in the United States.* New York: McGraw-Hill Co. 1933. I. Ch. VIII. Pp. 382-442

working under both public and private auspices. It embraces treatment received from doctors, dentists, nurses, opticians, physiotherapists, chiropodists, midwives, and medical social workers and also from osteopaths, chiropractors and Christian Science practitioners. It includes hospitalization both for acute conditions and institutional care for chronic diseases.

The amount of medical care needed is determined directly by the amount of illness that exists, so that questions of the volume of service must be related to the amount of sickness itself. The kind and amount of care required varies with the nature and severity of illness; some illnesses do not require the attention of professional persons and only a small part demand hospitalization. The kind and amount of care necessary for a particular type of illness changes from time to time with advances in the knowledge of the efficacies of particular kinds of treatment. Neither is the amount of sickness a constant; public health measures and higher living standards for at least part of the population have practically eliminated some diseases and the same future is conservatively predicted for others. It has been estimated, for example, that had the 1900 case rate for typhoid fever, tuberculosis, and diphtheria in Michigan persisted until 1931 there would have been 45,460 cases of those diseases in the latter year instead of the 8,387 that were actually reported.[44]

The amount of medical care received is a function of the amount of sickness and of the cost of treatment and the funds available to pay for it. Medical care is provided by professional persons in private practice and as free care in public and other clinics. The ability of practitioners and hospitals to provide free care is determined by income from pay patients and from gifts, subsidies, and endowments. Medical care of certain types is

[44] See Sinai, Nathan. *Report of the Committee on Survey of Medical Services and Health Agencies.* Lansing: Michigan State Medical Society. 1933. P. 6

also provided at nominal or no cost to the patient by governmental agencies through taxation. The amount of care derived from this source is dependent on government income and, ultimately, on the financial capacity of the community.

The amount of medical care received varies with family income. Presumably, those who are well-to-do receive nearly all of the care that present standards of treatment indicate, while substandard care increases as income decreases; but there is evidence that those who are always poor exceed those of the middle classes in the proportion of their illnesses that receive certain types of medical attention, notably hospital and visiting nurse care. During the recent depression, strenuous efforts were made in many places to provide medical relief for those who most needed it. There is some possibility that one of the effects of the depression was to sweep a large number of wage earning families into the lowest income group where they obtained more medical care than they had previously had.

A. Provision of Care

Attempts to estimate the volume of medical care received from statistics of the activities of professional persons and of institutions are imperfect because: (1) the data are not related to the amount of care needed, i.e., the amount of sickness experienced (2) most of the care is provided by private practitioners whose records are not readily available and (3) the data cannot be accurately classified according to the economic status of the persons to whom the care is given. The latter point is important because depression changes have probably not been distributed uniformly over all income groups in the population.

Services of Medical Practitioners

It is well known that the number of patients applying to private medical practitioners for treatment decreased during the depression. In the study of *Economic Aspects of the Practice of*

Medicine in California,[45] Dodd determined through questionnaires returned by physicians, that the volume of paid practice decreased by 9 per cent between 1929 and 1934.[46] The same questionnaires revealed sharp declines in the total volume of professional charges and income over the six-year period (among physicians, dentists, and osteopaths) although these declines may represent decreased collections or reductions of fees as well as declines in the number of patients. Leven found, from 6,000 questionnaires returned by physicians, a 10 per cent decline in collections and a 3.6 per cent decline in charges as early as 1930.[47] It was estimated by the Federal Emergency Relief Administration that about three-tenths of one per cent or 600 of the physicians in the country were receiving public relief in 1934,[48] and the proportion doubtless would have been greater were physicians not being paid by the government to provide care for the indigent. It would be significant to know whether there was an interruption in the trend toward medical specialization in an effort to secure more patients or whether an increasing proportion of physicians had abandoned private practice to take salaried positions in industry or government.

Dodd compared some of his data with those of the Medical Care Committee's study[49] and discovered that the time spent by dentists at their chairs decreased from 32 to 28 hours per week.[50]

[45] Dodd. *Economic Aspects of the Practice of Medicine in California. Op. cit.* (See note 10, p. 80)

[46] Part of the decline may be explained by the partial retirement from practice of older physicians during the period of observation. However, the depression probably retarded or reversed this normal movement so that the 9 per cent decrease is due almost entirely to lack of patients.

[47] Leven, Maurice. *The Incomes of Physicians.* Publication of the Committee on the Costs of Medical Care. No. 24. Chicago: University of Chicago Press. 1933. P. 77

[48] Unpublished data from the Survey of Occupational Characteristics of the Relief Population, Federal Emergency Relief Administration, Washington

[49] Falk, Klem and Sinai. *Op. cit.* (See note 34, p. 132)

[50] Dodd. *Op. cit.* Since Dodd's study was limited to California and since the Committee study covered several communities scattered over the country, there

It is well-known that nurses, particularly those not regularly attached to hospitals, were hard hit by the depression; it was estimated that 8,000 were unemployed in 1933.[51]

Yet, none of this evidence gives adequate assurance of an actual decrease in the total amount of care provided. Dodd's study indicates that the volume of free work in private practice increased by 46 per cent and free care in clinics or hospitals by 69 per cent during the 1929-34 period.[52] The evidence supplied by the physicians cooperating with the California study is supported by scattered information from other sources. Margaret L. Plumley made a study of the *Growth of Clinics in the United States*[53] in which she found that in 1931 there were probably from 35 million to 50 million visits[54] made to an estimated 7,000 clinics and out-patient departments operated for ambulatory cases.[55] In a more recent discussion Miss Plumley noted great increases in the demand for clinic service during the early part of the depression. During the first nine months of 1931 there was a 20 per cent increase over 1930 in the number of patients at the North End Clinic in Detroit although the fees collected per visit were only 40 per cent of 1929 income.[56] Similar evidence came from the Harper Hospital out-patient department

may have been differential factors in the return of questionnaires to the two investigators.

[51] See Haupt, Alma C. "Some New Phases in Public Health Nursing." *American Journal of Public Health.* 25:1350, No. 2. December 1935

[52] Dodd. *Op. cit.* These proportions, when compared with the 9 per cent decline in private pay care, should not be taken to mean that the *total* volume of work increased, for the numerical bases from which the percentages were calculated differ greatly from one another. A misinterpretation of the question on the part of some physicians prevented Dodd from estimating net changes in the actual number of patients treated.

[53] Chicago: Julius Rosenwald Fund. 1932

[54] *Ibid.* P. 35

[55] *Ibid.* P. 36

[56] See Plumley, Margaret L. "Out-patient Departments and Clinics Fight Depression." *The Modern Hospital.* 38:99, No. 2. February 1932

in Detroit and from clinics in other cities. In 1929 Harper Hospital collected some fee from 80 per cent of its clinic patients but from only 20 per cent in 1930. At the same time this clinic began to refuse chronic patients who could not benefit from treatment and patients who had a family physician, unless the physician would refer the patient.[57] In Chicago, between 1929 and 1933, the clinics of non-government hospitals caring for patients unable to pay a private physician doubled the total volume of service rendered and quadrupled the volume of free service.[58] That many of these visits were made by erstwhile patients of private medical practitioners is indicated by the efforts that clinics have been forced to make to identify and refuse admission to such patients as long as there is any possibility of their meeting the charges made by their regular physicians.[59]

Further indication that emergency measures have been taken to supply medical care is found in some of the work projects of the relief administrations. Work projects for bedside nursing care and disease prevention work employed more than 7,000 nurses in 39 states.[60] In Portland, Oregon a medical clinic provided treatment for 54,000 patients in less than one year.[61] Medical care, on a large scale, was provided for relief clients through Federal Emergency Relief Administration Regulation No. 7 providing several different plans for the distribution of medical service. In addition to providing much of the care needed they had the secondary effect of giving an income to a large number of private practitioners some of whom might otherwise have been forced to go on relief.

---

[57] *Ibid.* P. 99

[58] See Ropchan, Alexander. *Chicago Hospital and Clinic Survey 1934.* Reprint from *Bulletin of the American Hospital Association.* P. 8, April, 1935

[59] See Plumley. *Op. cit.* Pp. 99-101

[60] See "Works Progress Administration Health Projects Give Employment to 7,444 Women." P. 1 (See note 30, p. 131)

[61] *Ibid.* P. 2

Hospital Treatment

The decline in payment for hospital care and the increasing need for free treatment was apparent early in the depression. In 1932, Davis and Rorem published a series of essays on the subject, *The Crisis in Hospital Finance and Other Studies in Hospital Economics.*[62] Reduced family income made it impossible for patients to purchase medical care, while decreased hospital income from pay patients and voluntary contributions reduced the amount of free care that hospitals could give. Many who formerly would have been pay patients postponed surgical care or obtained home care when possible. The results of these tendencies, according to Davis and Rorem, were increasing demands for government hospital care and increases in out-patient service; increased fixed operating cost per patient and insufficient utilization of facilities for efficient and economical operation with consequent bankruptcies, particularly among proprietary hospitals where there were mortgages; competition among hospitals for patients and attempts to seek the goodwill of doctors; and requests for government subsidies.[63] Simultaneously with the publication of these studies, numerous articles testifying to decreasing income appeared in hospital publications.[64]

Quantitative evidence of some of these changes came to

---

[62] Davis, Michael M. and Rorem, C. Rufus. *The Crisis in Hospital Finance and Other Studies in Hospital Economics.* Chicago: University of Chicago Press. 1932

[63] *Ibid.* Pp. 3-10

[64] See especially "Here Are the Figures from Which Occupancy Chart Was Constructed" (editorial). *Hospital Management.* February 1933; Doane, Joseph C. "Increasing the Hospital's Income in Times of Economic Stress," *The Modern Hospital.* July 1931; Woods, Charles S. "The Adjustment of Operation of Hospitals to the Economic Depression." *Bulletin of the American Hospital Association.* April 1932; Parnall, Christopher G. "Recent Changes in the Economic Aspects of Hospital Management." *Bulletin of the American Hospital Association.* July 1932; MacEachern, Malcolm T. "Some Economic Problems Affecting Hospitals Today." *Western Hospital Review.* July 1932

light in the Chicago study of general hospitals which revealed that during 1933, 88 per cent of government hospital beds were occupied but only 50 per cent of all non-government beds; 35 per cent of the beds in small (fewer than 60 beds) non-governmental hospitals were occupied as compared with 48 per cent in 1929.[65] In the short space of two years, 1931 to 1933, 13 large hospitals experienced a 30 per cent decline in income from

TABLE XI

INDEXES OF NUMBERS OF PATIENTS ADMITTED AND AVERAGE DAILY PATIENT CENSUS FOR GOVERNMENT AND NON-GOVERNMENT HOSPITALS: 1931–1935[a]

(1931 = 100)

Year	Government		Non-Government	
	Patients Admitted	Average Census	Patients Admitted	Average Census
1931	100	100	100	100
1932	112	107	97	96
1933	118	110	92	89
1934	118	113	94	90
1935	125	119	102	97

[a] Computed from *Hospital Service in the United States, 1936*, reprinted from *Journal of the American Medical Association*, March 7, 1936, Table 1, Pp. 787–89

patients. Among 28 non-government hospitals a drop of 43 per cent in service to pay patients between 1929 and 1933 was reported, and an increase from 22 per cent to 29 per cent in the volume of free service. This increase was made possible only through government subsidies.[66]

Large scale but less detailed evidence is derived from *Hospital Service in the United States, 1936,* an analysis of statistics of hospitals registered by the American Medical Association.[67] These data make it possible to compute index numbers of "pa-

[65] Ropchan. *Op. cit.* Pp. 4, 5. (Note 58, p. 145)

[66] *Ibid.* P. 7

[67] Reprinted from *Journal of the American Medical Association.* March 7, 1936

tients admitted" and "average daily patient census" for the years 1931 through 1935 for government and non-government hospitals separately. The index numbers using 1931 as the base are shown in Table XI.

These indexes show clearly the decline in nongovernment hospital business between 1931 and 1933; not only were fewer patients admitted as the depression approached the trough but those who were admitted stayed for shorter periods of time.

TABLE XII

INDEXES OF NUMBERS OF PATIENTS ADMITTED AND AVERAGE DAILY PATIENT CENSUS FOR HOSPITALS BY TYPE OF HOSPITAL: 1931–1935[a]

(1931 = 100)

Year	All Hospitals[b]		General		Nervous and Mental		Tuberculosis	
	Patients Admitted	Average Census	Patients Admitted	Average Census	Patients[c] Admitted	Average Census	Patients Admitted	Average Census
1931	100	100	100	100	—	100	100	100
1932	102	104	100	103	—	106	116	106
1933	98	104	96	94	—	111	105	108
1934	100	107	100	96	—	114	102	107
1935	108	113	109	105	—	119	107	108

[a] Computed from *Hospital Service in the United States, 1936*, Table 2, Pp. 790–91

[b] Includes data for maternity, industrial, and other smaller groups of hospitals rendering specialized services, not shown as separate classes.

[c] Data not available for 1931, but the number admitted increased from 171,000 to 173,000 between 1932 and 1935.

The studies mentioned in the preceding paragraphs indicate that the changes would have been more pronounced were data available for 1930 and 1929. Indexes for government hospitals show increases in both the number of patients admitted and the average daily occupancy, but they cannot be taken at their face value for comparison with non-government hospital records. Government hospitals include many institutions for the care of nervous and mental diseases and tuberculosis as well as hospitals operated for veterans and in connection with prisons and other agencies, while nongovernment hospitals are rather largely restricted to the care of acute conditions. Hence, depression changes for ordinary hospitals are better shown if they are classified according to type as in Table XII.

These data indicate a fairly steady situation with respect to patients admitted to general hospitals although the index numbers for average census show that the period of hospitalization for acute conditions declined. Comparative changes between government and non-government hospitals providing general care could be seen much more clearly from an analysis of statistics of hospitals classified at once according to both type of service and source of funds. An analysis pushed back at least as far as 1929 would be significant.

Hospitalization for nervous and mental diseases, tuberculosis, and other chronic affections presents a problem quite different from that of general hospital care for surgery and acute sickness. Patients suffering from chronic ailments are hospitalized for longer periods and are generally unable to pay for their care. The community has assumed responsibility for much of the care of tuberculous and mental diseases in government institutions. In general, the depression was not accompanied by large rises in the number of patients admitted to these hospitals but there was a tendency for those hospitalized to remain longer. These conclusions agree with the findings of a 1933-34 survey of state hospitals by the National Committee for Mental Hygiene, which revealed only a moderate rise in admissions and re-admissions. Overcrowding, refusals of admissions, and the makeshift retention of mentally sick in jails reported by some states were explained by cessation of normal building programs, reduction of budgets, and increased difficulty of paroling patients because of economic conditions. Budgetary restrictions also appear to have curtailed the usual amount of therapeutic and related care, thereby delaying recoveries.[68]

[68] See Komora, Paul O. *State Hospitals in the Depression.* New York: The National Committee for Mental Hygiene. 1934. Pp. 1-5

Care for persons suffering from other chronic conditions has largely been neglected or left to the sporadic efforts of private charity.[69] Recently there has been a discernible tendency for government to assume responsibility for these conditions which, like tuberculosis and mental disease, seriously impair the employability of the worker and which are of such prolonged duration that only the economically independent can afford the necessary treatment and institutional care.

Manufacture of Medical Aids

Statistics of the manufacture of drugs, biological products, optical goods, surgical and orthopedic appliances, etc., may provide a rough index of the volume of these types of medical treatment. Such data are available in some detail[70] and analyses could be made for the predepression, depression, and recovery periods. If value of products produced is used for this purpose it will be necessary, of course, to eliminate the long-time trends and to allow for changes in the purchasing power of the dollar. Drug grinding and druggists' preparations, patented or proprietary medicines and compounds, optical goods, surgical and orthopedic appliances, dentists' supplies and equipment, and biologics for human use can be treated separately and differentials in production studied.

B. MEDICAL CARE AND ECONOMIC STATUS

Five major studies have been devoted entirely or in part to the question of income as a determinant in the amount of medical care received. All have been family canvasses and all relate care received to the incidence of illness. One of the five projects was completed just at the turning point between the periods of

---

[69] See Jarrett, Mary C. *The Care of the Chronically Ill in New York City.* Presented to The Hospital Survey for New York by the Committee on Chronic Illness of the Welfare Council of New York City. July 1936. II

[70] See *Biennial Census of Manufactures.* Washington: U. S. Bureau of the Census. Vols. for 1929, 1931, 1933

prosperity and depression; the other four were undertaken during the depression.

The Studies

The study made by the Committee on the Costs of Medical Care from schedules collected during the years 1928 through 1931 is said to represent a period of normal business activity as opposed to a period of either unusual prosperity or pronounced inactivity.[71] The study covered 39,000 individuals in 8,758 white families in 130 scattered communities who were visited four to six times over the course of a year. Records of illness and medical care were analyzed according to a sixfold classification of income ranging from under $1,200 to $10,000 and over.

The Health and Depression Study of the United States Public Health Service and the Milbank Memorial Fund considered sickness and medical care for a three-month period in the early spring of 1933 in relation to income history from 1929 through 1932 in white working class families in samples of ten localities. All data were obtained in one family visit. The results for medical care are based on 6,700 families with 29,000 individuals in seven large cities.[72] Family income was used in two ways—income in 1932 just prior to the canvass and change in family income from 1929 to 1932. The analysis also included a classification of families according to employment status of the wage earners.

Shortly after the completion of this project Margaret C. Klem undertook a study of medical care and costs in California

[71] Falk, I. S., Klem, M. C., and Sinai, N. *The Incidence of Illness and the Receipt and Costs of Medical Care among Representative Family Groups.* Publication No. 26 of the Committee on the Costs of Medical Care. Chicago: University of Chicago Press. 1933. P. 33

[72] See Perrott, G. St. J., Sydenstricker, E. and Collins, S. D. "Medical Care During the Depression." Milbank Memorial Fund *Quarterly.* Vol. 12. No. 2. April 1934

families[73] for the State Relief Administration. This project and the analysis of the data were carried out in a similar way to the Health and Depression Study. Klem's study included 18,500 members of 5,100 families in 14 urban and 7 rural places.

Dodd collected schedules for 21,000 families containing about 65,000 persons in 26 California counties by the combined methods of house-to-house canvass and mailed questionnaires. The project attempted to cover the entire income range.[74]

The fifth major undertaking in this field is the National Health Inventory which is currently being analyzed by the United States Public Health Service. Enumerators made one visit to each of 865,000 families in 92 cities and 21 rural counties during the winter of 1935 and 1936. An effort was made to obtain records of serious illnesses (seven consecutive days of disability or longer) and their medical care over a one-year period for a cross-section of the population of each community surveyed. Practically the only data available at the time of this writing are for one Northern city.[75]

The major findings with reference to the receipt of care among various income levels are presented below. In all cases the amount of care is related to the amount of illness. While the results of the several studies are fairly consistent, it is not safe to make close comparisons of one study with another, for there are too many differences in methods of enumeration, training of interviewers, period of observation, type of population covered, and definitions.

Care Received from Physicians.—Almost without exception the proportion of illness attended by physicians was smallest among members of families at the bottom of the economic scale

---

[73] Klem, Margaret C. *Medical Care and Costs in California Families in Relation to Economic Status.* San Francisco: State Relief Administration of California. 1935

[74] Dodd. *Op. cit.* (See note 10, p. 80)

[75] Perrott, G. St. J. and Holland, Dorothy F. "Chronic Disease and Gross Impairments in a Northern Industrial Community." *Journal of the American Medical Association.* Vol. 108. No. 22. May 29 1937

and increased as family income increased. According to the Committee survey of the general population, 54 per cent of the illnesses in families with incomes of less than $1,200 were attended by private practitioners outside of hospitals and clinics. In the five higher income groups, from 63 to 66 per cent of the illnesses were attended. The care of a specialist was obtained in 4 per cent of the low income family illnesses and for 26 per cent of the illnesses in families with incomes of $10,000 or more.[76] Dodd's study, also covering the general population, showed a range in attendance by physicians of from 44 per cent to 90 per cent as income progressed from the lowest to the highest categories.[77] The range in the Health and Depression Study extended from 50 to 58 per cent, but it must be remembered that the population enumerated was limited to those in working class families.[78] Klem's analysis of physicians' care for illnesses involving disability revealed attendance upon 59 per cent in the relief population and an increase to 84 per cent for illnesses in families with incomes of $3,000 or more. There were 2,800 calls per 1,000 cases of disabling illness in the lowest income group, and 5,200 calls per 1,000 cases in families having incomes of $3,000 or more per year.[79] The Health Inventory data for one city showed that during the year of observation there were between 11 and 13 physician's calls for each illness of three or more months of duration in the lower income brackets, and 21 calls per case among members of $2,000 income families. Home calls varied from 7 per case in the relief group to 20 at the upper extreme.[80]

Turning the attention to the receipt of care among groups classified according to economic status in both 1929 and 1932, it is discovered in the Health and Depression Study data that

[76] Falk, Klem and Sinai. *Op. cit.* P. 282

[77] Dodd. *Op. cit.*

[78] Perrott, Sydenstricker, and Collins. *Op. cit.* (Note 72. P. 151)

[79] Klem. *Op. cit.* P. 22. (Note 73. P. 151)

[80] Perrott and Holland. *Op. cit.* Table 7. (Note 75. P. 152)

the smallest ratio of physician's care to illness, 47 per cent, occurred in the group that declined from a comfortable to a poor status over the four-year period. The highest proportion, 59 per cent, was found among those who were comfortable in both years.[81] Klem's data showed the smallest proportion of illnesses attended by physicians, 52 per cent, in the group that had been on relief in both 1929 and 1933 and the highest proportion, 80 per cent, in the group that had remained comfortable throughout.[82] The proportions of medically attended illness among the other economic groups in both studies range between the extremes in what appears to be somewhat illogical order.[83]

Hospital Care.—In the lowest income group (under $1,200) of the Committee survey, 7.4 per cent of the illnesses received hospital care. In the $1,200-1,999 group the proportion was 6.6 per cent, and thereafter an upward progression reached 8.6 per cent for illnesses experienced by persons in the $10,000 family income group.[84] In the Health and Depression Study and in the investigation conducted by Miss Klem, both of which were restricted to working class families, the variation in the amount of hospital care received was inversely correlated with the variation in family income. Between 9 and 10 per cent of the illnesses in the two lowest income groups of Klem's study were hospitalized, whereas the proportions receiving that service in the remaining groups declined to 7.5 per cent among those with $3,000 incomes.[85] In the Health and Depression Study, the variation was almost exactly the same.[86] It does not appear that

[81] Perrott, G. St. J., Sydenstricker, E., and Collins, S. D. "Medical Care During the Depression." Milbank Memorial Fund *Quarterly.* 12:108. No. 2. April 1934

[82] Klem, Margaret C. *Medical Care and Costs in California Families in Relation to Economic Status.* San Francisco: State Relief Administration of California. 1935. P. 24.

[83] Perrott, Sydenstricker and Collins. *loc. cit.;* Klem. *loc. cit.*

[84] Falk, Klem and Sinai. *Op. cit* P. 282

[85] Klem. *Op. cit.* P. 25

[86] Perrott, Sydenstricker and Collins. *loc. cit.*

there is any real inconsistency between the findings of these two studies and those of the Committee. Free hospitalization doubtless explains the seemingly high proportions at the lower levels of the economic scale and the truncated income ranges of the two depression studies stopped short of the upper economic levels, where, according to the Committee results, hospital care is more common than it is at the lower levels.

Considering the receipt of hospital care in connection with change in economic status, it is found in the Health and Depression Study that the chronic poor were the most frequent recipients of this type of service. Next followed those groups that had suffered a decline in income and the families which had always been in moderate or comfortable circumstances received the smallest amount of hospital care.[87] Among the California families studied by Klem the larger proportion of hospitalized illnesses occurred among those families which had always been in poor or moderate circumstances and those which had recently come to poor estate. There was no consistent drop in the proportion of hospitalized cases until the comfortable income level was reached.[88]

Dental Care.—In both the Committee study and the research conducted by Dodd, from 10 to 20 per cent of the individuals in the lowest income groups had received dental service during the year of observation, as compared with 60 to 90 per cent of the individuals in the highest income groups.[89] Klem found that 26 per cent of those in the lowest, and 45 per cent in the highest income groups had had dental service. One-fifth of the individuals in the relief population had never received dental treatment, but only one-tenth of those in the higher income ranges were completely without dental experience.[90] In an analy-

[87] *Ibid.* P. 108
[88] Klem. *Op. cit.* P. 86
[89] Falk, Klem and Sinai. *Op. cit.* P. 101; Dodd. *Op. cit.*
[90] Klem. *Op. cit.* Pp. 32, 33

sis of data collected in one city in connection with the National Health Inventory, Britten found that among white servants, unskilled and semi-skilled workers only from 16 to 18 per cent had received dental care, exclusive of extractions only, during a year; whereas from 30 to 40 per cent of those in the white collar occupations had received care.[91]

Nursing Care.—The Committee included all nursing service regardless of whether it was obtained from public or private sources, with the result that the highest proportions of illnesses attended by nurses, more than 10 per cent, occurred in the lowest income group where free care was obtained.[92] The Health and Depression Study and the Klem project limited their data to the receipt of care by visiting nurses. Both analyses indicated that the greatest proportion of nursing care of this type was received by the lowest income groups with a steady decline as income increased.[93]

Care Received from Secondary and Non-Medical Practitioners.—Only the Committee study collected data with reference to care received from osteopaths, chiropractors, Christian Science and other faith healers, midwives, and less definitely identified groups. At the lowest income level 1.7 per cent of the illnesses were attended by such a practitioner, while the proportions rose to nearly 7 per cent for illness at the $10,000 level.[94] This classification of practitioners doubtless covers too much ground, for midwives are employed more extensively in low than in high income groups.

Summary.—On the basis of the studies reviewed, there is unmistakable evidence of less medical care in the lower income brackets. While upon first examination the data of the several

---

[91] Britten, Rollo H. "Dental Care in a Large Northern City." To appear in *Public Health Reports* in the near future

[92] Falk, Klem and Sinai. *Op. cit.* P. 282

[93] Perrott, Sydenstricker and Collins. *loc. cit.*

[94] Falk, Klem and Sinai. *loc. cit.*

projects do not always appear to be in close agreement, study of the conclusions in the light of variation in definitions, periods and populations covered, and other factors result in the disappearance of most of the discrepancies.

c. Studies Suggested

(1) Study of Hospital Trends from Data Compiled by the American Medical Association.—The annual report of hospital statistics compiled by the American Medical Association was discussed earlier.[95] It is suggested that the data for individual hospitals shown in this report be tabulated to show trends of patients cared for by type of hospital for government and non-government hospitals separately. Both the number of patients admitted and the average census should be studied according to this detailed classification. Special hospitals such as those for maternity care, children, orthopedic conditions, etc., should be considered separately in this analysis. These studies can be carried on in such a way as to show differentials by region and size of community.

(2) Hospital Trends as Revealed by the National Health Inventory.—One of the Health Inventory projects involved a transcription of hospital data covering class of work, number of beds, admissions, average census, and type of control for alternate years from 1928 to 1936 from files of the American College of Surgeons and the American Hospital Association. Analysis of these data according to kind of work done and type of control (government or non-government) will be significant from the point of view of trends.

(3) Trends in Out-Patient Work as Shown by the National Health Inventory.—Another Health Inventory project collected information from out-patient departments covering type of

[95] "Hospital Service in the United States, 1936." Reprinted from *Journal of the American Medical Association.* March 7 1936

control, kinds of work done, income and expenditures, number of visits, personnel, and salaries. These data are limited to the year 1935, but should afford a base against which to make comparisons during years later in the recovery period.

(4) Trends in Patients Admitted and Length of Stay in Hospitals for Mental Disease.—The data published by the American Medical Association (see No. 1 above) provide a basis for studying the number of patients admitted and the average length of stay in hospitals for the care of mentally diseased and tuberculous patients. Reference was made in the discussion to possible delays in granting paroles to persons with mental illnesses and to reduction in the amount of treatment. The data from the American Medical Association report should be augmented with information collected directly from hospitals concerning amount of expenditures per case, reasons for failure to parole, amount of treatment rendered, and possible restrictions in the normal building program.

(5) Other Hospital Studies.—Studies of changes in the amount of hospitalization are suggested in Chapter II, see Pp. 54-55, Studies (4) and (8). These might be expanded to show the number and proportion of persons receiving care in private and semi-private rooms as opposed to the number and proportion receiving care in wards.

(6) Local Studies of Hospital and Clinic Work.—Both the Pittsburgh Health Department study and the Chicago Study of Hospitals and Clinics indicate that it is feasible, by intensive local studies, to obtain from the records of some public and private clinics and hospitals the total amount of care given, the amount that is fee and non-pay, and the amount of payment per patient over a period of years. Studies of these items should be extended to other cities and should separate relief clients from those who are only medically indigent.

(7) Amount of Nursing Care Provided.—There are various

methods of arriving at some estimate of the amount of nursing care provided. Hospital records will show the number of patients having private or special nurses during the period of hospitalization and the number of patient days. The records of registries for nurses may yield some information with reference to the number of nurses obtaining work outside of hospitals and the amount of work enjoyed. Records of visiting nurse associations and of health departments should show fluctuations in the number employed over a period of years.

(8) Births Attended by Midwives.—Perhaps there was an increase in the number of births attended by midwives. If properly filled out, a birth certificate should show the kind of attendant and a tabulation of this item for years before, during, and after the depression should be worthwhile.

(9) Maternity Home Care.—The *Social Statistics Bulletin* of the United States Children's Bureau publishes data on the number of persons cared for in maternity homes. These data can be analyzed for the entire depression and recovery period, but they should be studied in relation to the changes in the number of births.

(10) Free Care Provided by Emergency Agencies.—The Federal Emergency Relief Administration and Works Progress Administration made elaborate efforts to provide care by physicians and nurses for those who were unable to make direct purchases. Although the records are, on the whole, extremely poor, it should be possible, in individual communities to determine the amount of service rendered under these programs. Some data, such as the number of nurses employed on work projects to provide bedside care, should be available from the statistics collected by the Federal Emergency Relief Administration and Works Progress Administration for the entire country.

(11) Recanvass of Health and Depression Study Areas to Determine Changes in Amount of Medical Care Received.—

The real effects of the depression on the amount of medical care received can be shown best from family studies when the data are classified according to income history. It is impossible to obtain from families, historical records of ordinary sickness and medical care over an extended period of time. However, as was suggested in Chapter II, a repetition of the Health and Depression Study would seem worthwhile if families can be classified in such a way as to identify the group that had been unemployed and poor but later had been re-employed and were again in moderate or comfortable circumstances. The amount of medical care received by this group in comparison with the amount of illness should lead to fairly safe inferences when seen in the light of rates of care for those who have had jobs and have been comfortable throughout the depression and for those who were comfortable before the depression but became unemployed and are still unemployed. The suggested recanvass should include questions with reference to whether the items of care were or will be paid for.

(12) Effect of the Depression on Hospitalization of Confinement Cases.—Birth certificates show whether births occurred in hospitals. Changes in hospitalization of maternity cases can be determined from these records. Possible differentials by economic status can be inferred from the occupation of the father. If such a study is undertaken, it should be remembered that regulations prohibited the use of federal funds for hospitalization of relief clients, but that physicians were granted $15 for each home delivery; this would probably tend to discourage some hospital cases.

(13) Changes in the Purchase of Medical Equipment.—The adequacy of service rendered by hospitals, physicians, and dentists is determined to some extent by the possession of adequate and modern equipment. It would probably be possible to collect from a sample of hospitals, physicians, and dentists, records of

equipment purchases over a period of years; some of these data with reference to physicians and dentists might be available from income tax returns. Manufacturing firms could also furnish good data for major items of equipment. It has been suggested previously in this chapter that data on the manufacture of surgical appliances, drugs, biologics, etc., can be obtained from the *Census of Manufactures.*

(14) Manufacture of Hearing Aids.—Hearing aids for those who are deaf are manufactured by a limited number of establishments. It would probably not be difficult to obtain from them records of the annual production of hearing units.

ORGANIZATION FOR MEDICAL CARE

The organization for the practice of medicine, both preventive and therapeutic, may have undergone extensive changes during the depression. Temporary changes, such as those in the incomes of physicians, increase of free clinic care, or subsidizing of practitioners and hospitals, are of interest because of the possibility of their having accelerated previously established trends or initiated new movements in the direction of providing adequate care for a larger proportion of the population. The private practice of medicine was severely hit by the depression, but the community stepped in with a larger program than the one ordinarily conducted and changes were instituted which may have permanent aspects.

A. Medical Practice and the Depression

Earlier paragraphs have made it clear that there are no comprehensive statistics of increased demands for free medical care, but, if five million families applied for public assistance to carry on the normal process of living, it is safe to assume that those among them who experienced sickness were unable to pay the usual charges for treatment. Evidence already cited from several sources indicates large increases in services given by free clinics and decreases in service purchases from private practitioners. This marked change in the effective demand, i.e., the demand accompanied by ability to pay for medical and dental care, must have had serious implications for the professions and for non-government hospitals.

Implications for Professional Persons and for Non-Government Hospitals

Practioners.—The United States Department of Commerce recently prepared a bulletin showing the decline in professional

TABLE XIII

AVERAGE NET INCOME IN 1929 AND INDEXES OF INCOME FROM PRACTICE OF PHYSICIANS, DENTISTS, CONSULTING ENGINEERS, AND PUBLIC ACCOUNTANTS, 1929–1932[a]

(1929 = 100)

Year	Physicians	Dentists	Consulting Engineers	Public Accountants	B.L.S. Cost of Living Index
No. Cases	2,263	1,333	330	618	—
1929: income	$5,602	$5,199	$5,678	$12,846	—
1929: index	100	100	100	100	100
1930	95	96	84	94	97
1931	81	83	50	76	89
1932	61	61	28	62	80

[a] *National Income, 1929–32.* Senate Document No. 124. 73rd Congress, 2nd Session. Washington: Government Printing Office. 1934

incomes between 1929 and 1932.[96] The data are given in Table XIII in comparison with the Index of the Cost of Living computed by the United States Bureau of Labor Statistics.

While these data cover relatively small samples (there are more than 160,000 physicians in the country and more than 70,000 dentists)[97] Dodd has parallel evidence from larger samples of

[96] *National Income, 1929-32.* Senate Document No. 124, 73rd Congress, 2nd Session. Washington: Government Printing Office. 1934

[97] *American Medical Directory, 1936.* Chicago: American Medical Association.

California practitioners as does Leven from a sample of 6,000.[98]

Hospitals.—Non-government hospitals have had similar experiences. There are testimonies but few statistics on the reduction of pay patients and increased demands for ward care. Computations based on the annual report of *Hospital Service in the*

TABLE XIV
INDEXES OF NUMBERS OF HOSPITALS AND OF HOSPITAL BEDS, 1931–1935[a]
(1931 = 100)

Year	Hospitals		Beds	
	Government	Private	Government	Private
1931	100	100	100	100
1932	99	99	106	101
1933	98	97	108	100
1934	96	96	112	99
1935	95	94	116	100

[a] Computed from *Hospital Service in the United States, 1936.* Pp. 787–789

United States, 1936[99] show a recent decline in the number of hospitals, both government and nongovernment, but in terms of the number of beds nongovernment hospitals remained stationary while federal, state, and local government hospitals increased 16 per cent between 1931 and 1935 as seen in Table XIV.

That the increase in bed capacity of government hospitals was not accounted for by hospitals for mental disease and tuberculosis but was, at least in part, for general care is evident from the same source.[100] Data collected by the National Health Inventory will provide additional evidence of hospital changes in facilities, personnel, and expenditures over the period 1928-36.

1936; *Fifteenth Census of the United States, 1930. Population.* IV. Washington: Bureau of the Census. Table 3, p. 14

[98] Dodd. *Op. cit.* (See note 10, p. 80); Leven, Maurice. *The Incomes of Physicians.* Publication of the Committee on the Costs of Medical Care. No. 24. Chicago: University of Chicago Press. 1933. P. 77

[99] Reprinted from *Journal of the American Medical Association.* March 7 1936

[100] *Ibid.* Pp. 790, 791

Efforts to Accommodate

It is known that medical practitioners and hospitals attempted to meet the depression by lowering fees, that spreading costs over a period of time by giving credit was adopted by some, while others spread medical charges by adopting the insurance principle. Impetus given to insurance would probably have the effect of increasing the number of persons making medical payments.

Increased practice may have been sought by changing location or by a tendency of young men just entering practice to go to towns or rural areas. In Michigan, the 1910-20 trend of physicians into Detroit was reversed between 1920 and 1930 and the reversal seemed about to take on increased momentum in the 1929-31 period.[101] The *American Medical Directory*[102] provides data for a study of mobility of physicians over a period of time and for determining whether there has been a shift in the size of town chosen by new graduates for establishing a practice. Any tendency of the latter sort would probably make for a better quality of medical care for the populations of those places.[103]

Increased practice may also be secured by adding a general practice to a field of specialization or by complete abandonment of specialization. Recent studies[104] indicate that from 16 to 40 per cent of physicians are normally full or complete specialists. There are data in the *American Medical Directory* and

[101] Sinai, Nathan. *Report of the Committee on Survey of Medical Services and Health Agencies.* Lansing: Michigan State Medical Society. 1933. P. 51

[102] 1936

[103] Sinai is of the opinion that the customary practice of allocating rural population to that of places under 2,500 for the purpose of calculating the ratio of physicians to population is not universally accurate. Hence, in a Michigan study he distributed the rural population evenly among the cities of different population classes, and obtained a higher ratio of physicians in places of under 2,500 than in larger communities. See *Report of the Committee on Survey of Medical Services and Health Agencies.* P. 56

[104] See *Ibid.* P. 50; and Leland, R. G. *Distribution of Physicians in the United States.* Chicago: American Medical Association. 1935

elsewhere that may be employed to determine whether there has been a reversal or a letup of the long-time trend toward specialization.

There is probably an increasing number of positions open to physicians, dentists, and nurses in private industry and government Whether physicians have taken advantage of the opportunities to an increasing extent during the depression, may be determined from a study of the information about each physician recorded in the *Directory*. Analysis of *Directory* data would also reveal the possible retirement or changes to other occupations of an increasing number of physicians if they found it undesirable to follow a decreasing practice. It is not unlikely, however, that others who had retired were forced to make reëntries by reason of investment losses. Statistics of government agencies would show changes in employment of medically trained persons.

Another occupational adjustment might be indicated by the number of young men entering medical school. There are only 77 accepted schools of medicine in the United States[105] and it should not be difficult to discover possible fluctuations in the number of applicants for predepression, depression, and postdepression years. Since there is a lag of from 6 to 10 years between the time when the individual decides to enter medicine or dentistry and the time he is ready for practice, possible changes in the supply of practitioners due to this factor have probably not yet begun to appear. A by-product of such an effect of the depression might be an increase of such practitioners as osteopaths, chiropractors, and faith healers whose education takes less time.

Hospitals find it difficult to convert their equipment and buildings to other uses and are forced to effect adjustments through other means. With the decrease in income from pay patients and from gifts and endowments, non-government hospitals

[105] *Hospital Service in the United States, 1936.* P. 78

found it necessary to restrict the number of free and part-pay patients accepted.[106] One writer for the *Bulletin of the American Hospital Association*[107] enumerated 18 different ways in which hospitals could reduce expenses to meet the depression. Other writers suggested the same and other reductions.[108]

B. The Community Takes a Hand

When emergency funds were used to give medical care to the indigent, government was not taking an unheralded step. Long ago society established certain disease prevention activities as government functions and has increasingly assumed responsibility for making other preventive and treatment services available to at least some of those who could not pay for them. Everyday examples of the latter are immunization service, medical care for the indigent, and hospitalization of the tuberculous and insane. Changes did take place during the depression, however; some of them apparently being temporary in nature and others being set up on a long-time basis. Regardless of their nature they should be studied because they may possess new techniques that will be useful in future depressions or because they may have a permanent effect on medical practice.

Emergency Measures.—The provision of medical relief took a number of forms, the most frequent of which was direct payments, from emergency funds, to private agencies and practitioners. Many state and local relief agencies paid nongovernment hospitals out of local funds for care rendered to relief

---

[106] See Davis, Michael M. and Rorem, C. Rufus. *The Crisis in Hospital Finance and Other Studies in Hospital Economics.* Chicago: University of Chicago Press. 1932

[107] Woods, Charles S. "The Adjustment of Operation of Hospitals to the Economic Depression." *Bulletin of the American Hospital Association.* April 1932

[108] MacEachern, Malcolm T. "Some Economic Problems Affecting Hospitals Today." *Western Hospital Review.* July 1932; Parnall, Christopher G. "Recent Changes in the Economic Aspects of Hospital Management." *Bulletin of the American Hospital Association.* July 1932

clients. Federal funds were used to pay private physicians and dentists for treatment of relief clients outside of hospitals. The scale of fees was stipulated, cooperating practitioners were registered and were required to secure approval for treatment from the relief agency or a committee formed for that purpose. This plan was used in more than 30 states. Medical and dental care was given for acute illnesses interfering with earning capacity, endangering life, or threatening permanent preventable handicap. Nursing service was provided when, in the judgment of the physician, it was warranted.[109] In other places, where city or county physicians were already a part of the government health organization, their services were extended or used in conjunction with other forms of medical relief. Under the Works Progress Administration program, those who needed medical care were expected to pay for it from their wages or secure the assistance of local welfare agencies. Some services, chiefly those that could be given by nurses, were supplied through work relief projects which also gave work to unemployed nurses and other professional persons. Statistics of the number of examinations, special tests, immunizations, and home visits made under the work relief program were cited in another section. It is felt by some that these large scale medical assistance programs may have caused many people to become habituated to looking to public agencies for medical care. Studies should be initiated to determine whether this has been the case.

Federal and local emergency funds were used also for building and improving hospitals, clinics, and sanitaria. As of September 1936, there had been almost 400 such projects with estimated expenditures of more than $40,000,000.[110]

Permanent Developments.—What appear to be permanent developments are found in the extension of former programs

[109] Perrott, G. St. J. *Medical Care Under the Federal Emergency Relief Administration.* An unpublished memorandum in the files of the United States Public Health Service. 1934

[110] From the records of the Federal Works Progress Administration

and the beginning of new ones under the aegis of the Social Security Act; growth of group hospitalization and extension of credit, and of private group clinics; establishment of medical relief on a permanent basis; and the beginning of cooperative agencies for providing medical care.

The Social Security Act makes several provisions for health work, among them being expansion of facilities for care of crippled children; provision of maternity nursing service in rural areas; organization of maternal, infant, preschool, and nutrition conferences; postgraduate training for obstetricians, pediatricians, public health nurses, and midwives; advanced training at five universities for special groups of public health workers selected by their superiors;[111] industrial hygiene activities; health education for the public; and many other activities.[112] Records are already being made by the Children's Bureau and the Public Health Service of the advances and accomplishments being made.

Association of physicians in private group clinics for the purpose of rendering all types of services and sharing expensive equipment is not a depression phenomenon. The effects of the depression on them would be significant to know, however, for nearly all are willing to extend credit; economies of operation permit reduction of fees below the ordinary scale; and they readily lend themselves to provision of complete annual service for groups of patients. Study might reveal that the depression had brought new patients to them or that, because of its widespread nature, it had caused major reductions in the volume of work. Rorem's study of 55 clinics in 1931 provides a basis for future investigations.[113]

[111] "Public Health Training at Ann Arbor Under the Social Security Act," *The Health Officer*. United States Public Health Service. Vol. 1. No. 5. September 1936

[112] Eliot, Martha M. "Progress in Maternal and Child Welfare Under the Social Security Act." *American Journal of Public Health*. Vol. 26. No. 12. December 1936

[113] Rorem, C. Rufus. *Private Group Clinics, the Administrative and Economic*

Group hospitalization is a newer development. At the end of 1932 four or five hospitals had started arrangements with groups of people through which hospital care could be paid for in advance for about $10 a year. The American Hospital Association encouraged the plan and by the end of 1936 group hospitalization had been established in 60 cities and 600,000 members were eligible for benefits.[114] Group hospitalization has been furthered by the Canadian Hospital Council which has stimulated the development in 32 centers in 6 provinces.[115] The movement, on the whole, took great strides during the depression, but individual units may have suffered from inability of persons to maintain their memberships.

In Ontario the plan of giving medical relief through direct government subsidization of professional persons is being placed on a permanent basis, and elaborate accounting records have been set up which will make it possible to effect refinements in the system when they appear to be needed.[116] A similar plan for giving certain preventive services has been in operation in Detroit for nearly a decade.

Perhaps the newest of all plans, at least on a large scale, is that of the Resettlement Administration for providing medical, dental, surgical, hospital, and nursing service on a cooperative basis. Relief families in the Resettlement Administration program in North and South Dakota are paying fixed annual fees for care. The plan was inaugurated in these states during May 1937 and is to be extended to include non-relief agricultural workers. A similar plan was incorporated into the statutes of

Aspects of Group Medical Practice, etc. Publication of the Committee on the Costs of Medical Care. No. 8. Chicago: University of Chicago Press. 1931

[114] Davis, Michael M. "Next Moves in Medical Care," *Survey Graphic.* 26:70-72. No. 2. February 1937

[115] "Encouraging Group Hospitaliation." *The Health Officer.* U. S. Public Health Service. 1:397. No. 10. February 1937

[116] *Medical Relief Administration. The Experience in Essex County, Ontario,* Windsor: Essex County Medical Economic Research. 1937

British Columbia during the past year, but has not yet been placed in operation.[117]

C. Summary

Public health work and the practice of medicine are undergoing marked changes. Because many of them took place during the depression and because they affect health so vitally they should be studied intensively.

D. Studies Suggested

A large body of data available for conducting research in this field is found in the successive issues of the *American Medical Directory*.[118] Editions of the *Directory* were published in 1936, 1934, 1931, 1929, 1927, and for nine earlier years commencing with 1906. Presumably there have been additions to the list of items now collected about each physician as well as improvements in the technique of collection and verification; nevertheless, there is here a source of valuable information. Extended effort is made to include every licensed practicing and retired doctor of medicine in the country (and in Canada, as well) and to give recent information concerning his membership in the American Medical Association; date of birth; color; place in which his office is located; name of his medical school and date of graduation; date he obtained a license to practice; specialty, if he reports any, and whether he confines his practice to a specialty; membership in societies relating to specialized practice; and whether in active practice, engaged in administration, laboratory work, teaching, or retired. Names are classified alphabetically by city and for the entire United States. The information is obtained through questionnaires circulated annually and

[117] *A Plan of Health Insurance for British Columbia*, Victoria, B.C. Department of the Provincial Secretary. 1935

[118] 1936

verified through information secured from state license bureaus, medical schools, and local medical societies. The *Directory* for 1936 contains approximately 165,000 names.

Some of the studies that may be made from successive editions of this *Directory* are:

(1) Distribution of Physicians by Size of Community.—Changes in the distribution of physicians by size of place in which the office is located. To throw some light on changing trends among physicians just entering practice and on the significance for rural and small-town populations, this study can be conducted with reference to the age and length of time in practice, color of the practitioner, and by geographic area.

(2) Mobility of Physicians.—Mobility of individual physicians in relation to size of city or town in which the practice is located and by age and years in practice. The alphabetical listing for the country as a whole makes it a simple matter to follow a particular individual from one edition to the next although he may have moved his practice.

(3) Retirements, Withdrawals from Practice, and Reëntries.—Retirements, withdrawals from private practice for other reasons and reëntries in predepression, depression, and recovery years.

(4) Changes in the Number of Specialists.—Shifts in the number and proportion reporting specialization, limitation of practice to the specialty, and membership in societies of specialists. This study would suffer from the limitation that there are many degrees of specialization and hence, there may be some lack of objectivity in the reports. Furthermore, physicians who widened their practices during the depression may have failed to report that fact because they expected to resume limited practice upon the return of normal economic conditions or for other reasons. A student undertaking a study of this problem should be familiar with Sinai's study which revealed differences in the proportion of specialists when data were obtained from

(a) physicians themselves and (b) other physicians practicing in the same locality.[119]

(5) Proportion of Recent Graduates Entering Private Practice and Number of Medical School Applicants.—Trend in number of physicians entering private practice upon completion of training. This trend should be compared with the number of applications for permission to enter medical schools, which could be obtained from records of the schools. Use of applicants rather than students is suggested because the number of admissions is believed to be almost constant.

The studies proposed from the *Directory* data can be made for either a sample of the listings chosen at random or for all physicians. If the entire group is included the analysis would be facilitated through use of mechanical tabulation; if only a small sample is employed the desired data could be transferred to small cards and sorted by hand.

Studies that may be made from other data are:

(6) Relief Agency Records of Kinds and Amounts of Care Given.—Records of relief agencies or committees for the administration of medical relief will yield information in a few localities with reference to the kinds and amount of care given to public welfare recipients and the number of recipients involved. Similarly, records of city and county physicians' offices and of public hospitals can be used to show fluctuations in the volume and kinds of services rendered and in the number of patients treated. Either study could probably be related to the total number of persons on relief and hence eligible for care. A study of this nature has just been completed by Sinai and others from data obtained in several Michigan counties.[120]

---

[119] Sinai. *Op. cit.* Pp. 57-64. (See note 101, p. 164)

[120] Sinai, Nathan; Hall, Marguerite; Hoge, V. M. and Steep, Miriam. *Medical Relief in Michigan,* a study of medical relief in 10 counties currently planned for publication as a Public Health Bulletin of the U. S. Public Health Service in the near future

(7) Free and Pay Care in Non-Government Hospitals.—Non-government hospital records can be analyzed to reveal depression changes in pay and private room care as compared with free and ward care. They would also show the extent of governmentally subsidized care among hospitals that participated in the relief program.

(8) Changes in Number of Hospitals, Bed Capacity, and Occupancy.—The National Health Inventory is just completing an analysis of data showing changes in hospital occupancy and finances between 1928 and 1936. The data were obtained from the files of the American Medical Association, the American Hospital Association, and the American College of Surgeons.

(9) Evaluation of Systems for Giving Medical Relief.—Emergency programs for medical care of persons receiving relief should be evaluated in terms of: (1) the extent to which they served the public need (2) their cost to the community and (3) the degree of economic and other satisfaction they provided the practitioners. Perrott has already said, tentatively, that many programs were often inadequate because they generally failed to provide hospitalization, surgical care, care for chronic tuberculous and venereal disease cases, glasses and other appliances, and preventive services.[121] Some data for covering these points are matters of record and the machinery is already in motion to make some studies. In Ontario the physician makes appropriate entries on the left-hand side of a punch card for each patient treated and forwards the cards at monthly intervals to a central tabulating unit. The same plan was initiated in South Dakota during May 1937.

(10) Hospital and Professional Fees and Use of Credit.—Questionnaires to hospitals, physicians and dentists, if sponsored by an organization with sufficient prestige, would yield valuable information on changes in professional and hospital fees and the use of credit in meeting charges.

[121] Perrott. *Op. cit.* (See note 109, p. 167)

(11) Group Hospitalization.—The growth of group hospitalization should be studied and the records used to show changes in the number of persons participating. Studies of this nature can be made on either a local or a national basis.

(12) Distribution of Physicians, Dentists, Nurses, and Other Practitioners from Census Data.—The 1930 census schedule can be used to determine the number of physicians, dentists, osteopaths, nurses, and midwives in every county of the United States. From the viewpoint of availability of medical service these data should be made available for comparison with the 1940 Census.

Chapter V

Summary

HEALTH is a personally and socially desirable attribute because of its significance in relation to the ability of people to work and engage in other activities of living. Large amounts of money are spent and a complex organization of institutions and professional people has been created to preserve health and to care for those who become ill. Hence, a program to determine the social effects of depressions must necessarily give attention to changes that may have occurred in the health of the population and in the environment and institutions related to health.

Health and the receipt of medical care are associated with economic status. The poor have more sickness than the well-to-do and, in general, receive less medical care and live in an environment that is less desirable from the standpoint of health. The association between poverty and sickness is explained in two ways: (1) the conditions of poverty seem to produce ill health and (2) illness that incapacitates a worker often forces him below the poverty level.

What the recent depression may have done to health is not known. Since unemployment tends to select the unfit, one effect may have been to concentrate on the relief rolls many of those who did not enjoy good health without increasing the number of sick persons at all. However, in view of the association mentioned above, it may be supposed that widespread loss of income would have brought about an increase in the amount of sickness and death. On the other hand, many of the healthy people who were reduced to a condition of poverty may have possessed resourcefulness, initiative, and intelligence that enabled them to guard against sickness during the emergency. Or, perhaps insufficient time has elapsed since its onset for the depression to have had a noticeable effect on health. None of these questions has been answered, although they are of great importance.

Lack of income, itself, does not cause illness; the true causes, in so far as their origin is social, appear to lie in insufficient and improper diet, poor housing, hazardous working conditions, inadequate clothing, and other factors closely related to income. While research has thus far failed to reveal much of the process by which these circumstances give rise to ill health, it is believed that they are of sufficient importance to merit study in the present connection.

Services for the protection of health and the care of illness, also part of the environment, are singled out for separate consideration since they exist only because of the desire of the population to maintain itself in a state of well-being. It is certain that the depression increased the number of families unable to purchase medical services and that this occurred just when governmental and private health agencies found it most difficult to obtain funds for operating expenses. In the distribution of emergency funds it was necessary to provide for health as well as other needs. It is still not known whether more or less care was received during the depression, whether emergency health measures offset the declining activities of regular agencies, and whether the organization for medical care underwent changes that are likely to endure and alter the future health of the population.

The purpose of this memorandum is to call attention to changes that may have taken place and to suggest research in health, in the environment which seems to influence health, in the receipt of and organization for the provision of medical care. As in other fields of social science the research problem is not easy: measures of health are imperfect; records of sickness are incomplete and often not centralized; health agencies are numerous and diversified; record keeping for health work is just beginning to develop; and many studies depend on canvasses of families and other persons, in which method forgetting plays an important rôle. Of these difficulties the student is warned, as well as of the fact that in a field as untried as this one many undertakings will probably yield negative results. This would be no new phenomenon, however, and the significance of the subject appears to warrant undertaking researches that are grouped under the three topics mentioned above.

Measures of Health

Under this title the discussion concerns infant, neo-natal, and maternal mortality and stillbirths; general mortality, particularly of males of working ages, and mortality from accidents and suicide; morbidity from sickness among members of sick benefit associations, school children, and the general population; accidents; and nutrition.

Infant and Maternal Mortality, etc.—One group of studies would look for trends in infant, neo-natal, and maternal death and stillbirth rates over a period of 10 or 15 years by matching birth and death certificates. Through this matching the data could be classified by social-economic status of the father's occupation because the birth certificate carries that information if it is properly filled out. In the case of maternal mortality it might be possible to show whether the death occurred following a live birth, stillbirth, miscarriage, or abortion. It is also proposed that similar rates be calculated for groups of census tracts or other areas in large cities over a number of years covering the predepression, depression, and postdepression periods. Economic status of the area would be estimated from median rental data available for the tracts from the Census Bureau. It would be imperative to allocate births to the area of parents' residence.

General Mortality.—It is suggested that a careful study be made of mortality from various causes for specific age and sex groups with consideration given to depression or postdepression deviations from the long time trend. Similar studies are proposed for deaths due to accidents and suicide; the latter being classified by economic status. In this connection a special study is proposed because of its value for future research rather than because it has significance for the present problem. Occupations are often poorly reported on death certificates and it is difficult to calculate reliable death rates based on occupations reported for the total population. Hence, it is proposed that certificates for deaths occurring during the year or two immediately following the census of 1930 and 1940 be matched with the census returns for 1930 and 1940, respectively, to determine the extent to which the entries for occupation are comparable.

Morbidity.—While it is difficult to obtain sickness records for minor ailments over a period of years to note whether there were changes during the depression, it is believed that certain inferences may be obtained from a recanvass of the Health and Depression Study areas for the sake of comparing current illness rates among those who were unemployed but who became reemployed after the depression with rates among those who have been comfortable throughout the depression and those who were comfortable before the depression but became unemployed and are still unemployed. Records of insurance companies might yield valuable data if the selective effect of dropping policies which was probably greater among some economic groups than others, could be eliminated. Attendance records of school children are usually well kept and sometimes include data for calculating sickness rates over a period of years according to the occupation of the father.

Hospital records are a useful source of morbidity data if they can be made available. It would be significant to study trends before, during, and after the depression in the number of hospital and clinic cases, especially by cause of illness, if all hospitals and clinics in a city agree to cooperate. Comparisons of case fatality rates during the depression cycle between the poor and the well-to-do are also important. It would seem worthwhile to look for changes in the number of patients admitted to hospitals for mental disease and to continue efforts already under way to determine the extent of nonhospitalized cases among the low and the upper income groups.

Histories of the frequency of abortions, stillbirths, and miscarriages for the depression period can be obtained for women in different income groups if superior canvassers are available.

Accidents.—Data seem to be available from several sources for arriving at industrial accident rates through the period of depression and recovery. Rates would have to be calculated on the basis of hours of exposure rather than number of workers to avoid the bias of part-time work during the depression.

Nutrition, Height and Weight.—Nutritional status is hard to measure, but some conclusions may be derived from school records of height and weight, gains in weight, and physical examinations which are probably available for a period of several years. The subjects could be grouped according to economic status of the parents and according to different types of relief policies under which many of them were living. Weight of infants at birth also affords a basis for a possible study of changes correlated with those of the business cycle.

Environment and Health

Chapter III gives consideration to changes in living standards; consumption of food, clothing, and housing; and changes in the occupational environment. It is suggested that these matters be studied because they must be related to health although the present state of knowledge does not permit one to infer that changes are necessarily indicative of changes in health. The relationships are too complex for that.

Living Standards.—Using existing studies and data rather easily available, it is possible to establish a level below which income probably should not fall if families are to maintain good health. Additional data on income could be used to estimate the proportion of families that fell below the subsistence standard. Also, it is suggested that income from relief agencies be compared to the estimated budgetary requirements to determine the adequacy of relief.

Consumption.—The *Biennial Census of Manufactures* and other data can be used to estimate the production of foodstuff, clothing, plumbing materials, and other articles needed for healthful living. Similarly, records of city engineering and building departments and of real estate boards and other agencies can often be used to determine the number of families that were forced to double up during the depression and the amount of new building that took place.

Occupational Environment.—In the past, the provision and enforcement of safety codes and compensation laws has been correlated with the frequency of industrial accidents. It is proposed therefore that studies be made of changes in compensation laws and statutes regarding safety practices. Attention should be given to an increase or decrease in the stringency of regulations, number of workers covered and number of industries included, and in the number of diseases made compensable. Studies of changes in the frequency of inspections, in expenditures for inspection service, and in plant expenditures for the provision of safety devices might also yield significant data if they were extended over the depression and recovery period.

Accidents are also related to fatigue and the condition of machinery. Hence, studies of accidents are suggested in relation to changes in the length of the working day and working week and to the cause of the accidents such as failure of machine parts, etc.

The Prevention and Treatment of Illness

This section discusses changes in the volume of preventive services and in the amount of care provided for sickness, as well as changes in professional practice and in the organization for giving medical treatment.

Prevention.—Most important in this field is the need of a system for recording statistics of public health work and the designation of an agency for collecting and compiling such statistics. Specific studies are suggested for determining changes during and after the depression in health work, in terms of expenditures, personnel, and volume of service rendered. Studies can be made from data already collected in the Health Conservation Contests, by the National Health Inventory, by the Bureau of the Census in *Financial Statistics of Cities,* and also from data in the files of health departments. Statistics collected for the first time during recent years by the United States Public Health Service can be used as bases for future studies of the volume of preventive services.

Research is also suggested to determine changes in the amount of immunization work done. Data may be obtained from health department records in some places and the Communicable Disease Study of the National Health Inventory also provides material for a study covering a period of 19 years.

The Receipt of Care.—Studies of trends in hospital service can be made, and in some cases are already being made, from data collected by the American Medical Association and the National Health Inventory. These studies would show the number of agencies, personnel employed, expenditures, and the amount of work done. Similar studies could be carried on intensively in individual cities where hospital and clinic records permit. Studies should also be made of the activities of emergency agencies to determine whether there were increases in the amount of care provided corresponding to decreases in care purchased and provided from private funds.

Studies of these types are limited because they do not show accurately changes in the amount of care received by persons at particular income levels. It is proposed, therefore, that the Health and Depression Study areas be recanvassed in connection with a study of sickness already proposed, to determine the amount of medical care received by families which have been reemployed since the depression, in comparison with families which have not been reemployed and families which were employed throughout the depression.

Studies are also suggested to determine the proportion of births taking place in hospitals, proportion attended by midwives, and changes in the manufacture and consumption of surgical appliances and medical equipment.

Organization.—One group of studies is designed to show whether the depression brought about changes in the distribution of physicians by size of community, the number retiring from practice, the number of medical school applicants, and the practice of specialties.

Medical relief from emergency government funds was given on a large scale. Records are in poor shape but studies should be made wherever possible, not only to discover the kind and amount of care given but also to evaluate the system of giving it. Already, one plan is being adopted on a permanent basis in some places, and the advantages and defects of all plans should be known for use in future depressions.

Additional investigations would reveal the effect of the depression on professional and hospital fees and on systems of group medical practice and group hospitalization.

Index

Abortions, statistics, 106
Abortions and miscarriages, and maternal mortality, 16-17; determining maternal mortality rates, 16; stillbirths and, 18-19; suggested study, 56-57
Accident Facts, 57
Accident prevention, 103
Accidents, 28, 57-59, 70-73; mortality from, 28; study by Metropolitan Life Insurance Co., 28, 29; suggested studies, 29, 59, 103; statistics, 57; automobile, 58; industrial, 58, 103
ACH Index of Nutritional Status, 10n, 68
Age, and mortality, 26-27; as a factor in school sickness, 44
Agricultural Statistics, 79
Agriculture, Census of, 78
American College of Surgeons, 157
American Hospital Association, 157
American Medical Association, 147-48, 157, 158
American Medical Directory, 163-64, 170-72
Annual Mortality Statistics for the United States, 27
Australian Census, 33
Automobile accidents, 58

Baldwin, B. T., 9n
Benefit associations, *see* Sick benefit associations
Beney, M. Ada, 99n
Bigelow, G. H., 56n
Births, records of, 6, 19-22; infant mortality studies based on, 19-21
Bloomfield, J. J., 127n
Bolduan, C. F., 47
Boots and shoes, consumption of, 92-93
Borowski, A. J., ix
Britten, Rollo H., ix, 9, 30n, 87, 88n, 96n, 156
Brundage, Dean K., 40n, 41n, 50n
Brunet, W. M., 48n
Budgets, minimum quantity, 74-77, 92-96; clothing, 92-94; food, 94-96
Bunzel, B., 28n
Business activity and mortality, 27

California, studies of medical care and costs, 151-57
California Medical Society study, 80, 132, 152
Carter, W. E., 10n
Case, J. D., 45n, 52n
Case reporting, *see* Registration
Census, Australian, 33
Census, Irish, 32-33
Census, U.S. Decennial, 32-33, 163
Census of Agriculture, 78
Census of Manufactures, 90n, 92, 93, 180
Census of Patients in Hospitals for Mental Disease, 55
Census of Population and Unemployment, Michigan, 80
Census tracts, basis for studies, 23-25; infant mortality by, 23; comparability of, 25

Chapin, Charles V., 1
Chapin, F. Stuart, 109
Chicago, infant mortality study, 23; decline in hospital income, 145; free clinic service, 146-47
Child health conservation programs, 130
Chronic disease, social effects of, 55-56
Civil Works Administration, emergency health programs, 124-25; child health conservation, 130
Clague, Ewan, 84
Clark, T., 10n
Cleveland, infant mortality study, 23
Cleanliness, 115
Clingersmith, David, ix
Clinics, extent of services by, 128, 129-31; family income and, 131; demand for care, 144-45; development of, 144-45, 167-68; growth in the United States, 144-45
Clothing, consumption of, 92-94; shoes as index; quantity budgets for, 93
Collins, Selwyn D., 10n, 13n, 14n, 18n, 19n, 34n, 35n, 36n, 38n, 39n, 43n, 44n, 45n, 46n, 50n, 97n, 128n, 132n, 151n, 154n, 156n
Committees, *see under name of committee*, 5-9. Municipal Health Department Practice, committee on
Commodities, estimating supply of, 93-94
Communicable disease records, 6; *see also* Disease
Communicable Disease Study, 140
Community in health, 166-69
Compensation laws, 102
Confinement cases, 160
Conservation of health, *see* Prevention and treatment of illness
Consumer in the depression, 94
Consumption and health, 86-96; housing, 86-92; clothing, 92-94; food, 94-96
Consumption in the Depression, 77n, 86n, 94
Costs of Medical Care, Committee on, 34-35, 38-39, 128, 131-32, 143, 151, 152-57
Cummings, H. W., 48n

Davis, Michael M., ix, 146, 166n, 169n
Deaths, records of, 6; relation to age, 26; trends in depression rates, 29; *see also* Mortality
Decline in income, *see* Income
De Harts, Stanford, 98n
Dental care, 155-56
Dentists, practice of, 143
Detroit, housing, 92; Harper Hospital, 144-45; North End Clinic, 144
Diagnostic Clinic, Philadelphia, 65
Diehl, Harold S., 71, 104n
Diet, and pellagra, 73; of low-income groups, 94-95
Diseases, and economic status, 1, 39; determining amount of, 6; records of: house-to-house canvasses, 7; correlation between physical fitness and, 11; suggested studies, 29, 49-57; sickness rates, 32-57; income and, 39-40, 73, 85; sickness surveys: family, 32-40; industrial, 40-42; voluntary sickness insurance, 41-42; study of hospitalized, 46-47; nature and study of reportable, 47-48; study of venereal, 48-49; nutrition and, 59-60; institutions and organizations for, 113; control of, 114-15, 124-25, 130-31, 167; services for, 127-33; public and private services, 128; *see also* School sickness
Diseases, prevention and treatment of, 5, 113-74, 181-82; institutions for,

113; environmental control, 114-25; services for, 114-40; health education, 125-27; direct preventive services, 127-33; preventive services: and family income, 131-33; problems of research, 133-34; suggested studies, 134-40, 157-61; receipt of medical care, 140-61, 181-82
Distribution of practitioners, 174
Doane, Joseph C., 146n
Dodd, Paul A., 80n, 132, 143, 144n, 152-53, 155, 163
Dorn, Harold F., 111n
Doull, J. A., 35n, 36n
Dreyer, G., 10n
Dublin, Louis I., 13n, 28n, 96n

Economic status, and health, 1-3, 38, 175-76; and school sickness and, 45-46; *see also* Diseases and economic status
Education, effects on health, 107-8; *see also* Health education
Edwards, M. S., 48n
Eliot, Martha M., 64, 68, 168n
Emergency care, 159
Emergency health conservation, 124-25, 130-31
Emergency health programs, *see* Health
Emmet, B., 42
Employers and the profit motive, 98-99
Environment, and health,. 73-112, 179-80; social, 105-12; control of, 114-25
Evans, C. C., 45n
Ewalt, Marian H., 122, 123n, 129n
Expenditures for health, 118-25, 127

Falk, I. S., 34n, 38n, 132n, 135, 143n, 151n, 153n-56n
Family, surveys of, as source of morbidity data, 32-40; organization and health, 106-7
Family in the Depression, The, 107
Family income, standards, 78; determining, 79-85; house-to-house canvasses, 79; preventive services and, 131-33
Fatigue among workers, 71
Federal Emergency Relief Administration, as a health agency, 116; child health conservation, 130; physicians receiving relief, 143; medical care, 145; *see also* Works Progress Administration
Federal Housing Administration as a health agency, 116
Federation of Social Agencies, Pittsburgh, 121
Fees, hospital and professional, 173
Ferrell, John A., 124n, 139
Financial Statistics of Cities, 181
Fisk, C. T., 98n
Flook, E. E., 123n, 124n, 139n
Fluidity of population, 109
Food, production during the depression: depression diets, 94; minimum requirements, 94-95
Foster, R. R., 89n
Fowler, E., 47n, 107n
Franzen, R., 10n, 68n
Free clinic service, Chicago, 145
Frost, W. H., 35n, 36n
Funds for health education, 127

Gafafer, W. M., 35n, 36n
Goldberger, Joseph, 2n, 33n, 73n
Goodrich, Carter, 88n
Gover, M., 35n, 36n
Government, control in industry, 98; non-government hospitals and, 147-50; *see also* United States
Green, H. W., 23
Group hospitalization, development,

169; suggested study, 174
Growth of clinics in the United States, 144-45
Hagerstown, study of school sickness, 43-44; school absences, 43; study of weights of children, 61-63
Hagerstown Morbidity Studies, 18n, 33n
Hall, Marguerite, 172n
Hamblen, A. D., 13n, 26n
Handbook of Labor Statistics, 57
Harmon, G. E., 45n
Harper Hospital, Detroit, 145
Hart, Hornell, 140
Haupt, Alma C., 144n
Hauser, Philip M., vi, 23
Health, economic status and, 1-3, 175-76; income, 2, 85; public, 3-4; measurements of, 3, 5-8, 9-72, 177-79; problems of measurement, 5; indexes of, 11; environment and, 73-112; housing, 86-88; depression effects on housing, 87-88; of workers, 97-105; family organization, 104-7; social environment, 105-15; effects on education and communication, 107-9; social work, 108-9; mobility, 109-12; conservation programs, 130-31; public health work, 134-40; rural trends, 138; emergency measures, 166-69; permanent developments in, 167-68; *see also* Diseases: School sickness
Health activities, budgetary declines in, 119-25; registration, 119, 134-37
Health agencies, measurement of services of, 114-25; multiplicity of, 115-16; depression effects on, 115-25, 129-31; urban expenditures, 119-23; statistics, 119, 134-37; rural, 123-24; emergency programs, 124-25, 130-31, 145; health education, 125-27; public and private, 128-31; local, 136-37
Health and Depression Studies, 39, 49, 65, 134, 151, 152-56; suggested restudy, 159
Health Conservation Contest, 121, 135-36
Health control, expenditures for, 119-23
Health education, 125-27; communication and, 107-8; influence of radio, 108; of press, 108, 139; agencies, 125-26; need for, 126; funds: Social Security Act, 127
Health environment, 3; control, 114-25; expenditures for control, 118-25
Health Facilities Study, 137
Health Insurance records, 41-42
Health Inventory, the National, ix, 37-38, 40, 51, 79, 137, 140, 152-56, 157, 163, 173
Health measures, emergency, 166-69
Health of the School Child, 44
Health personnel, in Pittsburgh, 123; declines in, 124
Health programs, depression effects on, 113-74
Health promotion, suggested study, 138
Hearing aids, 161
Hedrich, A. W., 48n
Height-weight-age tables, 9, 61-63, 76-79
Herman, W. B., 35n
Hiscock, Ira V., 45n, 52n, 122, 123n, 129n
Hoge, V. M., 172n
Holland, Dorothy F., 37n, 56n, 152n, 153n
Home, *see* Housing
Hookworm, 73
Hospital care, income decline from, 145-49; government and non-government, 147-49; indexes, 148-49; chronic illnesses and, 149; relation to income, 154; study of trends, 157-58
Hospital Service in the U.S., 147, 148,

157, 163, 165; Hospitalized cases, data, 46-47; nature of, 46; suggested studies, 53-54
Hospitals, decline in income, 145-49; indexes of service, 147-49; indexes of facilities, 163; depression effects, 163, 165; fees, 173; free and pay care in non-government, 173
Hours of work, 101, 103
House-to-house canvasses, 7
Housing, 86-92; construction and occupancy, 86-92; relation to health, 86-88; railway employees, 88; depression effects, 88-92; in Detroit, 92
Hygiene, industrial, 100, 127

Illegitimacy rate, 106
Illness, *see* Diseases: School sickness
Immunization, and income, 132; of children: study of trends, 140
Incidence type of family survey, 32-39
Income, relation to health, 2, 73, 85; history of, and sickness rates, 39-40; minimum requirements, 75-86; during depression, 77-80; below subsistence level, 77-81; of relief families, 81-86; of family: preventive services and, 131-33; receipt of care and, 131-33, 154-55, 156; of physicians, 142-43; of dentists, 143; nurses, 144; decline in, of professional persons, 142-44; decline in, of hospitals, 147-50
Industrial accidents, 57-59, 103-5
Industrial hygiene, 100, 127
Industry, technological changes, 99-100; hours, 101, 103
Infant mortality, *see* Mortality, infant
Institutional care, 148-50
Institutions for prevention of illness, 113, 115-17, 125-26
Insurance, study of records, 50-51; relation to unemployment, 50
. . . *Internal Migration in the Depression,* 109n
Irish Census, 32-33

Jacobs, Esther, 65
Jaffe, A. J., vi
Jarrett, Mary C., 56n, 150n
Jeans, Philip C., 68
Johns Hopkins School of Hygiene, 35

Keffer, R., 42n
King, Wilford I., 2n, 73n
Kiser, Clyde V., 35n, 65
Klem, Margaret C., 34n, 38n, 132n, 143n, 151-56
Kline, E. K., 35n
Komora, Paul O., 149n

Labor, hours of, 101, 103
Labor unions, activity, 99
Lazarsfeld, Paul F., *see* Stouffer, Samuel A. and Lazarsfeld, Paul F.
League of Nations, Health Organizations, measures of malnutrition, 67
Leland, R. G., 164n
Leven, Maurice, 77n, 143, 163
Lombard, H. L., 56n
Lorimer, Frank, 111n
Lotka, A. J., 13n
Low income, *see* Income

MacEachern, Malcolm T., 146n, 166n
McGill, Kenneth H., ix
Malaria in Russia, 110
Malnutrition, 59-72; and depression, 59-63, 64-65, 66; measures of, 61-65, 67-72; derivative effects, 69; suggested studies, 70-72; *see also* Nutrition
Manufactures, Census of, 90, 92, 93, 180
Marriages, 106

Mason, H. H., 45n
Mason, Robert C., 98n
Maternal mortality, *see* Mortality, maternal
Maternity care, standards, 132; suggested studies, 159
Mead, Pauline A., 124n, 139
Measures of health, *see* Health
Medical aids, changes in volume of manufacture, 150; suggested studies, 160
Medical care, scope of, 140-41; requirements, 140; relation to illness: determining amount needed, 141; decline in receipt of, 142-61; estimating receipt of: family income and, 142; provision of, 142-50; under FERA, 145; and economic status, 150-56, 175-77; from physicians, 152-54; organization for, 161-74; *see also* Hospital care
Medical Care, Committee on the Costs of, 34-35, 38-39, 128, 131-32, 143, 151, 152-57
Medical practice, and depression, 161-66; income from private, 162-63; community and, 166-69
Medical relief, 166-69; suggested study, 173
Medical schools, 165
Medical services of practitioners, 142-45; free and paid, 144-45; in Chicago, 145; *see also* Hospital care: Medical care
Mental cases, 54-55
Mental Hygiene . . . Committee, 149
Metropolitan Life Insurance Co., 28, 29
Michigan Census of Population and Unemployment, 80
Migration, *see* Mobility
Milbank Memorial Fund Studies, 35, *see also* Health and Depression Studies
Millar, William M., 107n
Minimum subsistence standards, 74-86; determination of, 75-76; rural-urban differences, 76; incomes below, 77-80; clothing, 92-93; food, 94-95
. . . *Minority Peoples in the Depression,* 109n
Miscarriages, *see* Abortions and miscarriages
Mobility, effects on health, 109-12; in the spread of malaria, 110; depression effects, 111
Moorhouse, G. W., 23n
Morbidity, *see* Diseases: School sickness
Mortality, 25-31; among taxpayers in Providence, 1; records of deaths, 6; as an index of physical fitness, 11-18; trends, 25-26; causes: relation to age, 26; business activity and, 27; from accidents: from suicide, 28; suggested studies, 29-31
Mortality, infant, 12, 13-16, 23-25; index of physical fitness, 11; records: rates, 12; major causes: "true infant mortality," 13; trends, 13-14; in England and Wales: family income and: study by Children's Bureau, 14; index of sanitary status, 15; occupation of father and, 19-21; suggested studies, 19-21, 23-25; census tracts in study of, 23-24
Mortality, maternal, 12, 16-18; rates, 16; Mortality, maternal, trends: diet and, 17; economic status and: in England, 17; suggested studies, 21-22, 25
Mortality, neo-natal, 11, 12, 13-16, 23n
Mortality, occupational, 28, 30-31, 96-102; rates for studies, 30; reports in England and Wales, 30-31; causes, 96; depression and, 96-97; employers' efforts: government control, 98; trade unions: technological changes, 99; hygiene and safety standards, 100-1; hours of work, 101

"Mortality in Certain States in Recent Years," 26
Moulton, Harold G., 77n
Mountin, Joseph W., ix, 123n, 124n, 135n, 139n
Municipal Health Department Practice, Committee on, 125n, 128n

National Committee for Mental Hygiene, 149
National Health Inventory, *see* Health Inventory, the National
National Income, 162
National Industrial Recovery Administration, 101
National Tuberculosis Association, 30
Neo-natal mortality, 11, 12, 13-16, 23n
Nesbit, O. B., 44n
News, The Detroit, 92
Newspapers and periodicals, suggested study, 108, 139
New York City health areas, 24
Non-medical care and income, 156
North End Clinic, Detroit, 144
Nurses, employment of, 130, 144, 167
Nursing care, 130, 167; and income, 156; suggested study, 158-59
Nutrition, 59-72; measures, 9-10, 60-72; Pirquet index, 10n; ACH index, 10n, 68; disease and, 59-60; income and, 60; depression effects, 60-63, 64-65, 66; weight as index, 60-64; physical examination, 64-66; relative value of indexes, 65-66; suggested studies, 70-72; *see also* Malnutrition

Occupational environment, 96-106, 127; changes, 97-100; depression and, 100-1; suggested studies, 102-5
Occupational groups in study of infant mortality, 19-21
Occupational mortality, *see* Mortality occupational
Ogburn, W. F., 27n
Ontario plan for medical relief, 169

Palmer, Carroll E., 39n, 61, 62
Palmer, G. T., 10n, 68, 69
Palmer, Gladys L., 105n
Palmer, Roy A., 98n
Parnall, Christopher G., 146n, 166n
Parran, Thomas, 4, 48
Peck, Gustav, 101n
Pellagra, 73
Pennell, E. H., 123n, 124n
Periodicals and newspapers, 108, 139
Perrott, G. St. J., ix, 2n, 37n, 38n, 39n, 50n, 56n, 151n-56n, 167n, 173n
Pfeiffer, A., 48n
Philadelphia, malnutrition in, 65
Physical examinations, physical fitness, 10; nutrition, 64-66; suggested studies, 70-72
Physical fitness, measures of, 9-11
Physical standards, validity of, 9
Physicians, records of institutions and, 6; income of, 142-43; number receiving relief, 143; care by: relation to income, 152-54; rural-urban distribution, 164; specialization of, 164-65; association in clinics, 168; suggested studies, 170-72; distribution of practitioners, 174
Pirquet, C., 10n
Pirquet system of nutrition, 10n
Pittsburgh, study of health budgets, 121; vaccinations, 129
Pittsburgh Health Department, 123
Plumley, Margaret L., ix, 144, 145n
Pneumonia, records of, 6
Portland, Ore., clinic care, 145
Poverty as a result of illness, 1
Powell, Webster, 84
Press, the, as a health agency, 108, 139
Prevalence type of family survey, 32-33

Prevention and treatment of illness, *see* Diseases, prevention and treatment of
Professional fees, 173
Professional income, decline in, 142-44
Profit motive and health of workers, 98
Providence, R. I., mortality among taxpayers, 1
Public health, 3-4, *see also* Health
Public health work, records: suggested studies, 134-40; rural trends, 138
Public Works Administration, suggested study of accidents, 104-5; as a health agency, 116

Quantity budgets, 74-76, 92-96; clothing, 92-94; food, 94-96
Queen, Stuart A., 109

Radio as a health agency, 108
Railway employees, living standards, 88-89
Randall, M. G., 35n
Reckless, W. C., 24n
Records, *see* Registration
Registrar General of Births, Deaths, and Marriages in England and Wales, 14, 15, 30
Registration, communicable diseases, 6; births and deaths, 6, 18-23; venereal disease, 6, 48-49; abortions and miscarriages, 16, 56-57; stillbirths, 18, 22-23; school sickness and absence, 42-43, 51-53; reportable diseases, 47-48; public health work, 117-19, 134-37
Relief, living standards and income of families on, 81-86; periods and phases, 82-83; direct, 83; records, 83-85; work, 84; *see also* Income
Relief agencies, records, 172; *see also* Civil Works Administration: Federal Emergency Relief Administration: Works Progress Administration
. . . *Relief Policies in the Depression,* 81n
Reportable diseases, nature of, 47; study, 47-48
Reporting, *see* Registration
Resettlement Administration health program, 169
Roberts, Lydia, 69
Rochester, Anna, 19n
Ropchan, Alexander, 145n, 147n
Rorem, C. Rufus, 146, 166n, 168n
Rural Electrification Administration as a health agency, 116
Rural health agencies, 123-24
. . . *Rural Life in the Depression,* 109n, 139n
Rural health services, 138
Rural-urban distribution of physicians, 164
Russian Congress on malaria, 110

Safety codes, 102-3
Safety promotion, 115
Safety standards in industry, 100-1, 102-3, 104
Sanderson, Dwight, . . . *Rural Life in the Depression,* 109n, 139n
Sanford, C. H., 45n
Sanitation and cleanliness, 115ff
Sayers, R. R., 127n
School absences, Hagerstown, Md., 43-44
School health programs, 139
School records, use of, 51-53, 139
School sickness, 42-46; studies, 43-46; Hagerstown study, 43-44; causes, 44-46; age, 44; economic status and, 45-46, 52-53; records: suggested study, 51-53 and grade: by district, 53
Shoes, consumption of, 92-94

Sick benefit associations, members, 40-42; records, 40-41; U. S. Public Health Service, study of, 40, 51
Sickness, *see* Diseases: School sickness
Sinai, Nathan, ix, 34n, 38n, 132n, 141n, 143n, 151n, 153n-56n, 164n, 172n
Smith, R. M., 45n
Social environment, relation to health, 106-12
Social Security Act, 127, 167-68
Social Statistics Bulletin, 129
Social work, effects on health, 108-9
Specialization of physicians, 164-65
Spencer, Lyle M., 106n
Standardization of Industrial Accident Statistics, 58
Standards, subsistence, 74-86; determination of, 75-77; incomes below, 76-81; rural-urban differences, 76; clothing, 92-94; food, 94-96
Statistical Abstract of the United States, 92
Stecker, Margaret L., 33n, 75, 87n, 93n, 94n
Steep, Miriam, 172n
Stiebeling, Hazel K., 94, 95
Stillbirths, 18-19, 22; certificates of, 22; suggested studies, 22, 25
Stix, Regine K., 18n, 65, 106n
Stouffer, Samuel A., vi, 106n; and Lazarsfeld, Paul F., . . . *Family in the Depression,* 107
Stouman, K., 135
Studies suggested, infant, neo-natal and maternal mortality and stillbirths, 19-25; general mortality, 29-32; accidents, 58-59; morbidity and sickness rates, 49-57; nutrition, height and weight, 66-72; family income and consumption, 77-95; occupational environment, 102-6; preventive services, 134-40; receipt of medical care, 157-61; organization for medical care, 170-74
Study of Consumer Purchases, 80
Suicide, 28-29; mortality from, 28-29; suggested study: study by Metropolitan Life Insurance Co., 29
Surgeon General of the United States, *see* Parran, Thomas
Surveys, family: incidence and prevalence types, 32-40
Syndenstricker, Edgar, vii, 2n, 4, 10, 14n, 18n, 26n, 27, 33, 35n, 48n, 73n, 96n, 151n, 153n-56n

Taussig, Fred J., 18n, 57n, 106n, 107n
Taylor, C. K., 10n
Taylor-Jones, L., 44n
Technological changes in industry, 99-100
Terborgh, George, 90
Thomas, D. S., 27n
Thompson, L. R., 9n
Thompson, Warren S., . . . *Internal Migration in the Depression,* 109n
Tobey, James A., 116n
Trade union activity, 99
Tropical malaria in Russia, 110
Truant officers' reports, 52
Truesdell, Leon E., 93
Tuberculosis, records of, 6
Tuberculosis Association, National, 30

Underweight, 61-64, *see also* Weight
United States:
Bureau of the Census, see Census
Bureau of Internal Revenue as a health agency, 116
Bureau of Labor Statistics: accident data, 57; minimum subsistence standards, 74-77; indexes of cost of living, 76-77, 162; hours of work, 101, 103; digest of workmen's compensation laws, 102
Bureau of Mines as a health agency, 116

Children's Bureau, as a health agency, 116; trends in receipt of care, 129-31; child health conservation, 130
Decennial Census, 32-33, 163
Department of Agriculture as a health agency, 116
Department of Commerce, as a health agency 116; data on professional incomes, 162-63
Department of Interior as a health agency, 116
Department of Labor as a health agency, 116
Office of Education as a health agency, 116
Public Health Service, ix; morbidity studies, 35; Health and Depression Studies, 39, 49, 65, 134, 151, 152-56, 159; study of benefit associations, 40, 51; quarantine regulation, 110-11; Domestic Quarantine Division health projects, 124-25; studies of health services, 128; Health Facilities Study 137; Communicable Disease Study, 140; *see also* Health Inventory, the National
Usilton, L. J., 48, 49n

Vaccinations, in Pittsburgh, 129; of children, 140
Vaile, Roland S., . . . *Consumption in the Depression,* 77n, 86n, 94
Vane, Robert J., Jr., 96n
Van Volkenburgh, V. A., 36n
Venereal disease, records, 6, 48-49; study of, 48-49
Voluntary sickness insurance, 41-42

Waller, C. E., 125n
Waples, Douglas, . . . *Reading in the Depression,* 107n
Warburton, Clark, 77n
Ward, Medora M., 94, 95
Warren, B. S., 99n
Watkins, J. H., 45n, 52n
Weight, physical fitness and, 9-10; nutrition and, 61-64; Hagerstown study, 61-63; depression and, 62-64; suggested studies, 69-70, 71-72
Wheeler, G. A., 2n, 33n, 73n
Whipple, G. C., 13n, 26n
White, R. Clyde and Mary K., . . . *Relief Policies in the Depression,* 81n
Whitman, G. E., 45n
Whitney, J. S., 30n
Wickens, David L., 89n
Wiehl, Dorothy G., ix, 33n, 35n, 39n, 60n, 81n
Williams, Lady, 17n
Wilson, C. C., 45n, 52n
Winslow, C.-E. A., 4
Wolman, Leo, 101n
Wood, Katherine D., 105n
Wood, T. D., 9n
Woodbury, R. M., 14, 15n
Woods, Charles S., 146n, 166n
Work relief, income from, 84; records, 84-85
Workers, health of, 97-105
Working hours in industry, 101, 103
Workmen's compensation, 102
Works Progress Administration: quantity budgets, 74-76, 92-96; subsistence standards, 75-77; suggested study, 105; as health agency, 116; health conservation, 124-25, 130-31; *see also* Federal Emergency Relief Administration
Woytinsky, Wladimir, 3n

Young, Donald, . . . *Minority Peoples in the Depression,* 109n

Zentmire, Zelma, 68

Studies in the Social Aspects of the Depression

AN ARNO PRESS/NEW YORK TIMES COLLECTION

Chapin, F. Stuart and Stuart A. Queen.
Research Memorandum on Social Work in the Depression. 1937.

Collins, Selwyn D. and Clark Tibbitts.
Research Memorandum on Social Aspects of Health in the Depression. 1937.

The Educational Policies Commission.
Research Memorandum on Education in the Depression. 1937.

Kincheloe, Samuel C.
Research Memorandum on Religion in the Depression. 1937.

Sanderson, Dwight.
Research Memorandum on Rural Life in the Depression. 1937.

Sellin, Thorsten.
Research Memorandum on Crime in the Depression. 1937.

Steiner, Jesse F.
Research Memorandum on Recreation in the Depression. 1937.

Stouffer, Samuel A. and Paul F. Lazarsfeld.
Research Memorandum on the Family in the Depression. 1937.

Thompson, Warren S.
Research Memorandum on Internal Migration in the Depression. 1937.

Vaile, Roland S.
Research Memorandum on Social Aspects of Consumption in the Depression. 1937.

Waples, Douglas.
Research Memorandum on Social Aspects of Reading in the Depression. 1937.

White, R. Clyde and Mary K. White.
Research Memorandum on Social Aspects of Relief Policies in the Depression. 1937.

Young, Donald.
Research Memorandum on Minority Peoples in the Depression. 1937.